The CAP and the Regions:
the Territorial Impact of the Common Agricultural Policy

The CAP and the Regions:
the Territorial Impact of the Common Agricultural Policy

Edited by

Mark Shucksmith
Kenneth J. Thomson
Deborah Roberts

EUROPEAN SPATIAL PLANNING
OBSERVATION NETWORK

CABI Publishing is a division of CAB International

CABI Publishing	CABI Publishing
CAB International	875 Massachusetts Avenue
Wallingford	7th Floor
Oxfordshire OX10 8DE	Cambridge, MA 02139
UK	USA
Tel: +44 (0)1491 832111	Tel: +1 617 395 4056
Fax: +44 (0)1491 833508	Fax: +1 617 354 6875
E-mail: cabi@cabi.org	E-mail: cabi-nao@cabi.org
Website: www.cabi-publishing.org	

A catalogue record for this book is available from the British Library, London, UK.

A catalogue record for this book is available from the Library of Congress, Washington, DC.

ISBN   0 85199 055 X
          978 0 85199 055 2

Printed and bound in the UK by Cromwell Press, Trowbridge, from copy supplied by the editors.

Contents

List of Tables

List of Figures

List of Maps

Preface

This book is the outcome of research project 2.1.3 "The Territorial Impact of the CAP and RDP" conducted within the framework of the ESPON 2000–2006 programme, partly financed through the INTERREG programme. The ESPON partnership consists of the EU Commission and the EU-25 Member States, plus Norway and Switzerland. The contents of this book do not necessarily reflect the opinion of the members of the ESPON Monitoring Committee. The project included the following partner institutions and personnel, whose contributions to the authorship of this book are gratefully acknowledged:

- **Arkleton Institute for Rural Development Research, University of Aberdeen (coordinator)**
  Dr Zografia Bika
  Dr Andrew Copus
  Ms Yvonne Loughrey
  Dr Deborah Roberts
  Dr Teresa Serra Sevesa
  Professor Mark Shucksmith (Project Leader)
  Professor Kenneth J. Thomson

- **Federal Institute for Less-Favoured and Mountain Areas, Wien**
  Dr Thomas Dax
  Ms Ingrid Machold
  Oliver Tamme

- **Institute of Spatial Planning, University of Dortmund**
  Professor Gunther Kroes
  Ms Martina Huelz

- **National Institute for Regional and Spatial Analysis, NUI Maynooth**
  Professor Jim Walsh
  Ms Jeanne Meldon
  Professor Patrick Commins

- **Norwegian Agricultural Economics Research Institute, Oslo** (joined April 2003)
  Dr Sjur Prestegard

- **Project Special Advisors**
  Dr Isabel Bardaji Azcárate (Universidad Politecnica de Madrid)
  Dr Tibor Ferencz (Budapest University of Economic Sciences)
  Dr Lars Olof Persson* (NordRegio, Stockholm).

*Lars Olof Persson died unexpectedly on 26 February 2005 after a short illness. He will be sorely missed by his ESPON colleagues with whom this book was written.

Acknowledgements

The editors of this book, and their research colleagues in ESPON project 2.1.3, would like to acknowledge the assistance of researchers in other ESPON projects during 2002–2004, and the staff of the ESPON Coordinating Unit. Staff of the European Commission also offered useful data and advice from time to time.

Preparation of this book benefited greatly from the expertise of Dr Andrew Copus as regards the maps, and of Mrs Alison Sage and Ms Deborah Murray as regards formatting.

Chapter 1

1. Introduction

1.1. Background

Agriculture occupies a central role in the economy, society and environment of the European continent, but is extremely diversified, geographically and structurally. These characteristics make the agricultural geography of Europe highly complex (Ilbery, 1981; Bowler, 1985; Robinson, 2004). Moreover, the sector has experienced many significant technological developments, and has been subject to a high degree of policy intervention, under national policies for a century or more. More recently, in an expanding European Union (EU), state management of farming has taken the form of the Common Agricultural Policy (CAP) while in the countries of Central and Eastern Europe, there has been more or less direct state control of farming.

The gradual coming together of most of the nation states of Europe within the EU has, perhaps paradoxically, increased the profile of its regions, as typified by the creation in 1992 of the Committee of the Regions as an ancillary EU institution, and more generally by increased cultural awareness within EU Member States of distinctive territorial characteristics, sometimes shared across national boundaries, or around the "periphery" of the EU. These characteristics are frequently agricultural and rural, and embody much of the rich historical and environmental heritage of Europe, distinguishing it from other "developed" continents such as North America and Australia.

However, economic forces within the EU have not always taken into account these spatially distinctive features, and indeed have sometimes acted to reduce territorial differences rather than to conserve and value them. In particular, transportation improvements, and the search for scale economies in manufacturing and services, have led to both congestion in urbanising regions and depopulation of more remote areas. While many of these economic forces are private, most transportation infrastructure is the responsibility of governments, at various levels from local and regional authorities to the European Commission's support for Trans-European Networks (TENs). Also, housing and other urban land uses have long been state-controlled. These considerations gave rise in the late 1980s to the initiative known as the European Spatial Development Perspective (ESDP), an attempt to achieve "the balanced and sustainable development of the territory of the EU" (Faludi and Waterhout, 2002). In 1999, EU-15 Ministers for Spatial Development and the Commission adopted the

ESDP, and in November 2004 EU-25 Ministers agreed "to focus their efforts …
on putting EU territorial issues and challenges on the political agenda, …
deepening the concept of territorial cohesion" (BMVBW, 2004).

However, chief amongst the policies of the EU has been the Common
Agricultural Policy, which for decades has occupied the major share of the EU
budget, and has had profound effects on farming in all regions, as well as
widespread economic, environmental, social, cultural and political implications.
Until recently, the effects of the CAP upon European regions were largely
ignored in the formulation of the Policy, except in terms of its Less-Favoured
Areas (LFA) regime for areas with natural handicaps. This attitude was a
reflection of the sectoral nature of the Policy itself, and of neglect by land-use
planners, who concentrated mainly on settlements, from megapolis to village,
rather than on the rural areas within which these are located. Indeed, study of
European rural economies in general, as opposed to farming itself, lagged far
behind that of urban economies, perhaps because the former show such wide
spatial diversity in comparison to the growing homogenisation of urban and civic
life. In recent years, however, the promotion of Rural Development Policy (RDP)
as "Pillar 2" of the CAP has involved greater attention to the regional aspects of
the Policy, and suggests some convergence with the concerns of the ESDP.

1.2. Aims and structure of this book

The EU's Second and Third Reports on Economic and Social Cohesion (EC,
2001c, 2004b) have called for the promotion of a more balanced and more
sustainable development of the European territory, in line with the ESDP, and
identified the need for further work on the territorial impacts of sectoral and
structural policies. "The Territorial Impact of the CAP/RDP" was therefore the
subject of a two-year project undertaken within the European Spatial Planning
Observatory Network (ESPON) 2000–2006 research programme (see
www.espon.lu). The overall aim of this study was to assess the extent to which
the CAP and RDP contribute to the goals and concepts of European spatial
development policies.

This book consists largely of the results of this project, whose aims were,
briefly, to use existing literature and data and to develop methods to analyse the
impact of the CAP on the regions of the EU (including New Member States, and
Norway and Switzerland). The study necessarily involved some collaboration
with other ESPON projects, but this book focuses on the core approach, methods
and conclusions of the CAP/RDP study itself. Given the major budgetary
resources and the many market interventions involved in the CAP, the main
numerical approach taken was largely econometric, with due attention paid to the
technological, geographical, socio-cultural and political forces also at work.

For reasons of time and space, the following pages cannot cover
descriptively the details of national, regional and local situations throughout
Europe. Rather, it is hoped that the book provides a general overview, and a

framework, for a greater appreciation of the territorial implications of a long-standing EU policy, in particular the conflicts with the objectives of the ESDP. In addition, an effort has been made to address the medium-term future, as regards both current CAP reforms being put in place for the period up to 2013, and farm household adjustments to these and other changes.

While general economic and other analyses of the CAP have been almost innumerable (see, for example, EC, 1994; Buckwell *et al.*, 1997; Ritson and Harvey, 1997; Fennell, 1997; Burrell and Oskam, 2000), previous studies of the territorial effects of the Policy have been relatively few and patchy. An earlier study carried out for the Commission (CEC, 1981) by the RICAP Group of agricultural economists was "beset by many difficulties of which the most important was without doubt the lack of centralised information both on trends in agricultural products and economic performance of regional agriculture within the Community" (CEC, 1981, p. 91). Confining itself to "descriptive analysis ... of the main regional agricultural trends" (often approximated by the authors themselves), it suggested that:

> (i) the agricultural systems of European regions [had] reacted markedly to the new characteristics of an agricultural common market progressively introduced by the CAP in proportion to their comparative advantages (natural, structural and technological factors);
> (ii) the restructuring of agricultural sectors brought about by favourable economic development plays an important part in the growth of regional incomes per agricultural worker; and
> (iii) during the period observed, the two mechanisms for increasing regional agricultural specialisation and restructuring agricultural sectors led to differentiated growth according to region thereby aggravating imbalances in regional agricultural incomes within the Community.

However, no specific policy instruments were analysed, nor were specific policy recommendations offered.

Bowler (1985) re-analysed the RICAP data for regional trends in the specialisation of eight crop and livestock products in 80 administrative regions, and concluded that "the process [of specialisation] has probably been accelerated by the security offered under the CAP", but was modified "according to the degree of market regulation offered to the various agricultural products of a region". He suggested that "the level of market support, and measures aimed at fostering structural changes in production, could be varied according to the particular character of agriculture in each of a defined set of regions". This would, however, have conflicted with the move towards a Single European Market, completed in 1992. More presciently, Bowler foresaw a mixture of reform of traditional CAP instruments, both evolution – e.g. strengthening the integrated development option – and devolution (though not of CAP price-setting, again for Single Market reasons) to national decision-making.

Little of this earlier work extended beyond farming itself. However, according to Robinson (2004, p. 72), some more recent CAP analysts, such as Ward (1993), Hoggart *et al.*, (1995) and Halfacree (1999), have considered the "territorialisation" of agricultural actor spaces, based on post-productivism (see below) and a wider consideration of future EU rural areas. These writers have generally based their work on a selection of EU regions chosen to illustrate their institutional arguments, rather than on a comprehensive analysis of both agricultural and non-agricultural data. Moreover, the different forms of emerging agricultures identified by these authors (e.g. Marsden, 2003) may be able to co-exist in close spatial proximity.

The structure of the book is as follows. After dealing with some basic concepts in the remainder of this chapter, Chapter 2 summarises territorial aspects of the rural areas and agricultural sector in the enlarged EU, and describes the CAP in some detail, emphasising its territorial elements, or lack of them. Chapter 3 outlines the relevant features of EU structural and cohesion policy and of the ESDP itself. These chapters provide essential background about the agricultural sector and the EU policies most relevant to European rural areas.

Chapter 4 examines numerically the territorial incidence of CAP and RDP measures and relates these through statistical analysis to the high-level ESDP objectives of social and economic cohesion, sustainability, and polycentricity. In Chapter 5, we investigate the topic by means of detailed case studies of a number of policy measures and countries. Chapter 6 uses further statistical analysis to estimate the territorial impacts of proposed reforms to the CAP and RDP in the context of enlargement. Chapter 7 returns to more conceptual analysis, by considering the CAP/RDP in the context of EU spatial policy, using the concepts introduced in Chapter 1. In the light of these findings, and key dimensions of rural development generally, Chapter 8 studies "good practice" in rural development policy, for example the LEADER approach.

Finally, after summarising the scientific findings of the research project, Chapter 9 presents the conclusions of the study, and offers some policy recommendations for improving agricultural and rural development policy in support of ESDP objectives.

1.3. Conceptual framework

A number of theoretical and empirical concepts need to be understood when approaching the numerical and qualitative analysis in later chapters. As stated above, the basic approach of the study is that of economics, including institutional economics (Commons, 1931; Williamson, 1996) alongside the basic axioms of production and market economics (Samuelson and Nordhaus, 2001). The merits or otherwise of this disciplinary approach to the territorial analysis of the CAP are left largely undiscussed here; alternatives such as those of environmental science or sociology might be considered, but seem more distant from the main considerations surrounding the actual development of the CAP and the ESDP. Nevertheless, the "political economy" of agriculture in developed countries needs to be recognised: in brief, the historical tendency to protect and support the farming sector (Ingersent and Rayner, 1999).

In addition to the economic perspective, two areas of conceptual consideration immediately present themselves: the empirical definition of "rural areas" in Europe, and the various concepts embedded in the ESDP – primarily that of "polycentricity" – and in analysis of CAP development. This section explores each of these areas in turn.

1.3.1.   Rural areas and rural economies

The conceptual and empirical definitions of "rural" are a perennial source of debate and experimentation, in both academic and administrative circles. For the present study, it was decided to use primarily the territorial scheme developed at the beginning of the 1990s by the OECD[1] Group of the Council on Rural Development for the collection and presentation of sub-national data, in an attempt to derive additional insights on national situations from international comparability. Details are given in Chapter 4.

In general, what areas are defined as "predominantly rural" etc. depends on (i) the definition and size of the basic units (e.g. land areas such as hectares, or households or communities), and (ii) how many and which of these basic units are grouped as a region (e.g. a "city region" or a "doughnut" around a city) to be classified as "rural" or otherwise. Thus clarity as to the hierarchical levels of territorial detail is central to this conceptual approach towards a territorial typology, and only through the different levels can the complexity of rural problems in various national and regional contexts be appreciated.

According to the OECD approach, about one-third of the total OECD population live in rural communities, occupying over 95% of the territory. National shares differ considerably, ranging from a rural population of under 10% in the Netherlands and Belgium to about 60% in Finland, Norway and Turkey.

In a further step of analysis, the change in total employment may be used as a primary indicator of the prospects for regional development. By comparing the regional employment performance with the relevant national average, a further differentiation of regions is achieved by means of a simple split into *leading* and *lagging* regions within each of the above three OECD types. In this way, while rural regions generally have tended to lag behind national averages, it has been shown that there are many rural regions, distributed over several OECD countries, which show dynamic development. This tendency is especially relevant for many rural areas in countries with a low degree of rurality, such as Germany and the United Kingdom. Prosperous rural areas are also found in largely "rural" countries with wide areas of low population density, as in Canada, Australia, the USA and Finland. There, the regional change of employment in dynamic rural

---

[1] The Organisation for Economic Cooperation and Development, based in Paris, is a forum for 30 developed-economy countries, including 19 in the current EU, plus Norway and Switzerland. It discusses, develops and refines economic and social policies.

areas is more than 10 percentage points higher than the national average. However, these statistical indicators do not give information on social development within those areas. As much of the created employment may be in low-wage sectors or attributed to part-time or unfavourable (e.g. inflexible) working conditions, further analysis of the contents of the employment growth is needed for a thorough assessment.

Some in-depth analysis within regions underpins the critical role of small and medium-sized towns in rural development (e.g. Dax, 1999). However, with the experience of substantial shifts in territorial employment growth, the assumption that urban centres are the main defining element of territorial development may be challenged. Looking at regional employment developments case by case, one can identify "leading" regions within all three types of regions. The recognition of diverse development performance is particularly important for predominantly rural areas as it rejects the notion that the mere fact of being located in a rural context would quasi-automatically lead to economic decline:

> Comparisons of employment change just between the three types of region do not reveal, however, a sufficiently detailed picture of the territorial diversity in development dynamics. The actual territorial patterns of development can not be properly appreciated just by looking at average figures of types of regions. Austria provides a clear but not untypical example. ... there are many rural regions that have been much more successful in employment creation than others. In fact, also in other countries many (dynamic) rural regions, although not the majority, perform better than many of the more urbanised (but lagging) regions .... This indicates that territorial development performance – in this case, success and failure to create additional regional employment opportunities – is not strictly correlated with the degree of rurality or urbanisation. Rurality in itself is not a handicap. It is not synonymous with decline, as much as urbanity and agglomeration are not automatic guarantees for prosperous development (OECD 1996a, p. 53).

International comparative analysis of sub-national economic data reveals that changes in employment structures over the last decades have led to a shift of labour force from the primary and secondary sectors to the tertiary sector. The level of agricultural employment is now restricted even in most predominantly rural areas, where the shares of agricultural employment do not exceed 20% on average (OECD 1996a, p. 45ff.) Nevertheless, the majority of land use remains shaped by agriculture, which underscores the spatial relevance of agricultural structures and the tight link to the rural economy.

1.3.2.   Polycentricity

The polycentricity concept marks a paradigm shift in thinking about Europe's spatial and economic structure, challenging the core-periphery model which tended to focus on a dichotomy in which a prosperous, economically dynamic core zone was contrasted with an under-developed, geographically remote periphery. At EU (and European) level, the core has been variously defined as the

"European Megalopolis", the "Blue Banana", the "Golden Triangle" and the "Pentagon" (Davoudi, 2003). The core-periphery pattern has been the key influential perspective in European regional policy discourse for more than two decades, with considerable influence over mainstream policy targeting. For example, the Structural Funds Objective 1 and the Cohesion Fund, as well as specific measures addressing peripheral disadvantage, such as the TENs initiative, various telematics schemes, and the Northern Periphery Programme (Article 10), have taken their prime objective from this territorial concept (Copus, 2001).

Several definitions may be offered for the concept of polycentrism. The classic definition of *morphological* polycentricity is of a region consisting of more than two cities that are historically and politically independent (where hierarchical relations are either absent or weak) and that are in proximity to each other and have a functional relation and complementary role to each other (de Boe and Hanquet, 2004). However, polycentricity may be focused on *functional* economic or political networks (Antikainen *et al.*, 2003), recognising that spatial proximity alone is not a sufficient condition for polycentric urban development. The relational aspect of polycentricity is based on networks of flows and cooperations between urban areas at different scales. Considering the opportunities and potentials of the countryside as an integral part of regional development, the structure of intra-regional flows and relations is of increasing relevance. Polycentricity results from institutional (political) processes, based on voluntary cooperation, and structural (economic, functional) processes, arising from "spontaneous" spatial development (Nordregio, 2003b, p. 3).

With respect to a comparative analysis of polycentricity against earlier concepts for spatial development, Schindegger and Tatzberger (2002) have stressed the following main features of the concept:

- Dynamism, considering cities not only as supplying centres (the "central places" concept, or "Zentrale-Orte-Konzept" in German discussion) but rather as driving forces for the regions.
- Not only a model of well-balanced settlement structure but of functional networks of existing and developing institutions, for example in the area of education, health, culture, leisure time and services which are able to coordinate and cooperate, and concentrate their efforts to produce synergetic effects.
- The activation of endogenous regional potentials rather than top-down measures of adjustment such as financial assistance and appropriate infrastructure. In polycentric development, the emphasis is shifted towards encouraging regional specialisation that can help firms to compete in global markets (Davoudi, 2003).
- Application at several levels or scales, implying a hierarchical interrelation of functional structures between the different levels.

According to the ESDP, the goal of balanced competitiveness is to be promoted through the adoption of strategies for polycentric urban development and new types of rural-urban relations. The ESDP vision is described as follows:

> The creation of several dynamic zones of global economic integration, well distributed throughout the EU territory and comprising a network of internationally accessible metropolitan regions and their linked hinterland (towns, cities· and rural areas of varying sizes), will play a key role in improving spatial balance in Europe (EC, 1999, para. 20).

The ESDP suggests a shift towards a more balanced polycentric system which will "help to avoid further excessive economic and demographic concentration in the core area of the EU" (EC, 1999, para. 67). It also affirms that such a policy will more fully utilize the potential of all regions, and so enhance the overall competitiveness of the EU (and cohesion) within a global context. Polycentric development should not remain restricted to Europe's larger metropolitan areas because this would not be "in line with the tradition of maintaining the urban and rural diversity of Europe" (EC, 1999, para 71). The guiding principle is the concept of urban hierarchies that cut across the whole of the EU territory.

However, pursuit of the concept of balanced competitiveness involves inherent contradictions between the requirements for global EU competitiveness as set out in the Lisbon Strategy (see e.g. Council of the EU, 2003) and the desire for an EU that is more socially and spatially cohesive. Thus the basic concept is inherently political, and is open to varying interpretations at different geographical scales and in different parts of the EU.

Empirically, polycentric development can be applied at different territorial scales, using the NUTS system[2]:

- at the European or *macro* scale: several metropolitan regions as global integration zones instead of only one prosperous, economically dynamic core zone;
- at the transnational/national or meso scale, corresponding approximately to the NUTS2 level: enforcement of a polycentric system of metropolitan regions, city clusters and city networks as well as systems of cities including the corresponding rural areas and towns;
- and at the regional/local or micro scale, corresponding to the NUTS3 level: enforcement of networking and cooperation between small and medium-sized towns as engines for economic development in rural regions.

---

[2] The Nomenclature des Unités Territoriales Statistiques, or Nomenclature of Territorial Units for Statistics, has been developed for the countries of the European Economic Area. It contains 3 levels (although some countries have defined up to 5), with about 1600 NUTS3 regions in the EU-27 (about 1100 in the EU-15).

The regional/local scale seems to be the most appropriate one at which to explore the relationship between rural areas and the concept of polycentricity. In this connection, the ESDP states that "the small and medium-sized towns and their inter-dependencies form important hubs and links, especially for rural regions. In "problem" rural regions, only these towns are capable of offering infrastructure and services for economic activities in the region and easing access to the bigger labour markets. Towns in the countryside therefore require particular attention in the preparation of integrated rural development strategies" (EC, 1999, para 93).

While a polycentric approach to urban development at different spatial scales may offer the prospect of a more efficient and more effective way of harnessing the potentials of regions, including the potential of rural areas through new rural-urban relations, there are many situations where the concept may not be applicable (Davoudi, 2003). The polycentric approach may not be a feasible option in low-density rural areas and/or where the transport infrastructure is weak. Additionally, the building up of institutional networks which are an essential component of the polycentric model may be a particularly difficult challenge for economically weak regions, especially in relatively more remote rural areas, which are often lagging behind precisely because of their lack of associational structures.

1.3.3.   Policy considerations

The need for room for decision making attuned to actual local situations supplements the approach taken in the analysis of Functional Urban Areas (FUAs) focusing on inter-municipal cooperation (Nordregio, 2003b, p. 9, p. 12). However, a broader place-based strategic remit requires the integration of the countryside into regional strategic frameworks; otherwise there is a risk of the countryside remaining as residual space, even at the local level.

Spatial and functional interdependencies between urban and rural areas are not a recent phenomenon, although the complexity of their linkages and relationships has often been underestimated. The physical and functional boundaries of urban and rural areas are becoming ever more blurred, while simultaneously the interdependencies are becoming more complex and dynamic, containing structural and functional urban-rural flows of people, capital, goods, information, technology and lifestyles. Waste and pollution also shape the fortunes of the cities as well as the countryside (CURS, 2004).

Whilst rural communities may be facing distinct challenges, it is now increasingly acknowledged that such challenges cannot be addressed in isolation from their wider context when it comes to policy formulation and programming. The functional interrelationships of urban areas with their surrounding countryside and the need to move away from the compartmentalization of policies are particularly highlighted in the ESDP by reinforcing the notion that the linkages between urban and rural areas should be based on an integrated treatment of the city *and* countryside as functional and spatial entities with diverse relationships and interdependencies. The following types of urban-rural relationships were distinguished in the Study Programme on European Spatial Planning (SPESP; BBR, 2001):

- home-work relationships;
- central place relationships;
- relationships between metropolitan and urban centres in rural and intermediate areas;
- relationships between rural and urban enterprises;
- rural areas as consumption areas for urban dweller;
- rural areas as open spaces for urban areas;
- rural areas as carriers of urban infrastructure (including waste treatment);
- rural areas as suppliers of natural resources for urban areas.

As Shucksmith (2003) has argued, numerous writers have attempted to analyse and explain the forces underlying the evolution and reform of the CAP[3]. For example, Swinnen (2003) has observed that the need to finance EU enlargement has been a major driver of budgetary proposals and decisions, while others have also emphasised the pressure arising from WTO negotiations. Taking a longer-term perspective, Buller (2003) has argued that the shifting dynamics of EU agri-rural policy reform reflect both an internal evolution and a motivation to maintain and perpetuate existing national shares of the CAP budget. Conceptualising policy changes in terms of four competing paradigms which vary in dominance over time (interventionist, neo-liberal, rural development, and multifunctional), Buller argues that it is these which have led to a fragmentation of the CAP into pillars, a move away from common policies towards regional and national policies, a growing diversity of instruments, and new approaches directed towards the production of public goods. Lowe *et al.* (2002) illustrated these tendencies further by comparing the contrasting British and French approaches to the Second Pillar of the CAP, and identifying "three new aspects": subsidiarity, multifunctionality, and territoriality. They foresee the CAP evolving into a broad regulatory framework, within which Member States can operate an increasing range of discretionary support measures, directed towards territorial priorities of agricultural landscape maintenance, employment, and sustainable rural economies.

For Marsden (2003), recent CAP reforms have essentially been attempts to deal with the growing crises of legitimacy in the dominant agro-industrial model: "to keep in place the basic principles of the industrial system while at the same time highlighting a rational conception of food quality". In competition with this, he argues, an alternative post-productivist model of the countryside has been promoted in north-west Europe, particularly, in order to shape the countryside socially and morally "in ways which continue to make it attractive and lucrative to aspiring ex-urban groups". The contest between these two models (agro-

---

[3] See Tracy (1997) for an authoritative and detailed account of the period up until Agenda 2000.

industrial and post-productivist) is embodied in the internal contradictions of the Agenda 2000 CAP reform. One proposes an agro-industrial "race to the bottom" through expansion and intensification which will facilitate competitiveness in global markets. The other promotes the coping mechanisms needed for managing the "consumption countryside". Both these models "for the social management of rural nature" tend to marginalize nature, whether through the production process or through a highly materialist conception of the consumption process. Moreover, Marsden argues, "both have their own socio-spatial expressions. In many rural regions in Europe they overlap across rural space and affect change in dual ways," each relying on market and state governance structures to manage the unsustainable conditions which they create. These regions are not identified, but, of particular interest for this book, Marsden argues (p. 13) that it is in those regions least exploited by either the agro-industrial or the post-productivist model, i.e. "peripheral rural regions", that an emergent sustainable rural development model may instead hold out greater hope.

1.4. Assessment objectives

Although many of the findings are also relevant to a more narrow assessment of the CAP and RDP against its own goals and objectives, such as adequate farm income levels, agricultural productivity improvements, de-intensification, and possibly higher or adequate diversity (e.g. mixed farming), the central aim of this book is to assess whether the CAP and RDP contribute to the goals and concepts of European spatial development policies. Thus key questions are whether the CAP and RDP support the ESDP goals of:

- social and economic cohesion
- environmental sustainability
- more polycentric development.

Each of the issues above can be considered at three levels described above – macro, meso, and micro.

1.4.1 Social and economic cohesion

The adoption in July 1987 of the Single European Act, which included new objectives in relation to *economic and social cohesion*, paved the way for a more integrated EU approach to spatial development. The objectives of the Economic and Social Cohesion policies addressed via the Structural Funds are summarised in Chapter 3.

The Third Report on Economic and Social Cohesion, *A New Partnership for Cohesion* (EC, 2004b), has given much prominence to the concept of territorial cohesion which goes beyond the more restrictive notion of economic and social cohesion. In policy terms, the objective of territorial cohesion is defined as helping to achieve a more balanced development by reducing existing disparities, preventing territorial imbalances and by promoting greater coherence between both sectoral policies that have spatial impacts and regional policy. Territorial cohesion also seeks to improve territorial integration and to encourage cooperation between regions. In essence, territorial cohesion seeks to ensure that people should not be disadvantaged by wherever they happen to live or work in the Union. Spatial policy and spatial development strategies are critical to the promotion of territorial cohesion.

1.4.2   Environmental sustainability

EU spatial policy has explicit goals of promoting sustainable development, prudent management and protection of nature and cultural heritage. For most of the first 30 years of the CAP, the Policy had no explicit environmental objectives. The development of EU environmental policy over the same period was very gradual and was mostly guided by a mainly reactive approach. It is not surprising, therefore, that the productivist orientation of the CAP until the early 1990s, supported by increasing levels of intensification and specialisation, contributed to a wide variety of negative environmental impacts. These include reductions in biodiversity, degradation and erosion of soils, contamination and excessive abstraction levels of water resources, air pollution by ammonia and greenhouse gases, destruction of wildlife habitats, and significant alterations to many distinguishing features of the European rural landscape (Baldock *et al.*, 2002). The incidence of environmental damage due to late twentieth century farming practices is not confined to the EU nor indeed can they be ascribed as being even primarily due to the CAP *per se*.

During the socialist era in Central and Eastern Europe, agriculture and food production were promoted by government plans that paid little attention to the suitability of production systems to the local environment. The pursuit of objectives related to increased production resulted in more intensive land use practices involving greater applications of inorganic fertilisers, and development of extensive drainage and irrigation schemes. While the levels of reliance on inorganic inputs remained much less than in the EU area there is evidence of considerable environmental damage.

It is important to note that, in addition to the differences between the EU area and the Central and Eastern Europe regions, there are also significant differences between regions in the two parts of Europe in relation to outcomes from the interaction of agricultural and environmental policies. Such differences are associated with contrasts in the levels of resilience of local environmental factors, the scale of operations and the modernisation/productivist stage attained by agriculture in each region.

Since the early 1990s, the relationship between agriculture and environmental policies has changed significantly. On the one hand, the importance of promoting more environmentally friendly farming practices has been adopted as part of the CAP objectives, and indeed the elaboration of the European Model of Agriculture (Cardwell, 2004) with the concept of multifunctionality has identified new policy-relevant roles for farmers as custodians of many rural-based public goods. On the other hand, EU environmental policy is now guided by sustainability principles which place more emphasis on prevention supported by a comprehensive regulatory system and there is more explicit emphasis on integration between policy areas.

In Central and Eastern Europe, the reform programmes introduced following the change of political regimes in the early 1990s have resulted in a decline in the overall intensity of agriculture, with fewer livestock and reduced usage of inorganic fertilisers and pesticides. The currently widespread pattern of relatively low-input and more extensive farming systems provides an opportunity for the development of more environmentally sustainable agriculture. The SAPARD programme provided an opportunity for the New Member States to include in their plans measures to support agricultural production methods designed to protect the environment and maintain the countryside. According to the EEA (2004) report on Europe's environment, many countries have included such measures in their SAPARD programmes, but there have been considerable delays with implementation, and most countries have given higher priority to improving competitiveness of the agri-food sector than to agri-environmental measures.

1.4.3   Accessibility

There are three main reasons why "geography really does matter in determining economic growth and performance ..." (McCann and Shefer, 2004; Rietveld and Vickerman, 2004):

- Whilst costs and journey times have followed a downward trend, both business and private users have gradually increased their demand for travel and transport. Within the business sector, this has been particularly true of high-technology industries, and those which have introduced more complex logistics, associated with low inventories. Other more traditional and resource-based industries, together with services offering standardised "products" have been less affected.
- Every region has not enjoyed the benefits of new or improved travel/transport or communications infrastructure to the same extent. In central regions, congestion may neutralise the benefits, while the change for remote regions may take the form of the reduction of natural protection (the so-called "pump effect"), exposing local businesses to overwhelming competition from more central areas which benefit from agglomerative advantages.
- The introduction of information technology has in many cases increased rather than reduced associated demands for business travel and transport.

The shift away from policies addressing core-periphery issues towards strategies to promote polycentricity does not, of course, imply that peripherality is a spent force. During the 1990s, a number of public bodies came to conclusions about the impact of new transport and communication technologies which would now (in a "post dot-com" world) be considered over-optimistic. Thus, for instance, the Committee of the Regions stated in 1998 that "Advances in communications technologies will ... bring major changes in the siting and nature of economic activity... The ESDP rightly sees ICT as a means of overcoming the adverse impact of geographical remoteness on business start-ups" (COR, 1999). Similarly the Conference of Peripheral Maritime Regions affirmed that "The advent of information highways is one of the aspects that has raised greatest hopes in the peripheries. The entry into the century of the immaterial would at last make it possible to do away with disparities linked to geographic distance...." (CPMR, 1997). However, more recently a number of academics have voiced a more sombre assessment, typified by the statement that "talk of the 'death of distance'... is unmistakeably premature..." (Rietveld and Vickerman, 2004, p. 241).

The principal theme of EU transport policy of relevance here is the Trans-European Networks (TENs) programme initiated in the 1990s, and intended to support the Community objectives of competitiveness and cohesion. Interregional competitiveness is expected to be enhanced through cost reductions resulting from more efficient transport systems. TENs provide new links and improvements to some existing network sections, and will result in an improvement in both the quantity and quality of infrastructure. By extending the networks into peripheral regions, which are more heavily dependent on agriculture, it is anticipated that there will be greater convergence between core and peripheral regions and, therefore, greater cohesion.

Chapter 2

2. The Common Agricultural Policy

2.1. Characteristics and trends in European agriculture

Agriculture forms the basis of the European food supply chain. Socially, culturally and symbolically, farming occupies a unique role as a traditional "way of life", from which identity is derived. Environmentally, agriculture "remains a major source of pressure on the environment ... becoming even more intensive and specialized" (EEA, 2001), while also having a positive role in maintaining valued habitats and landscapes. Structurally, EU agriculture has become more capital-intensive (more machines and buildings), more large-scale (fewer, larger commercial farms), less self-sufficient (more purchased inputs), and more regulated (for subsidy administration, food safety, animal welfare, etc.).

FAO-recorded agricultural land within the European countries covered by this book[4] occupies about 400 million hectares, i.e. about a quarter of the total land area, the rest being occupied by forest, rock and glaciers, urban construction, etc. Of this farmland, about half is arable (i.e. cropped), and the rest is either permanent pasture or under permanent crops (fruit orchards, vineyards, olive groves, etc.). The definition of "rural" and "agricultural" population (and even land coverage) is not simple, but in Europe as a whole, FAO records about a quarter of the total population as "rural", and about 8% as "agricultural". The economic significance of the European agricultural industry in terms of GDP share is smaller than this, due to lower labour productivity in farming, amalgamation with fishing, forestry and hunting, and a degree of under-recording[5], but lies between 2 and 3%.

Nevertheless, agriculture (along with imported products) forms the basis of the European food chain, which ranges from the supply of farming inputs (animal feed, agri-chemicals, machinery, hired labour) to the processing and distribution (including exporting) of food, drink and other farm-based products[6]. Food occupies a still-significant share of consumer expenditure even in highly

---

[4] Substantial agricultural resources and activities exist within excluded countries in Europe, i.e. Belarus, the western part of the Russian Federation, the countries of the former Republic of Yugoslavia, Ukraine, and a number of smaller countries, most of minor agricultural importance.

[5] It should be noted that much service employment and income is also under-recorded, being in the 'black' or 'grey' economies.

[6] For example, textiles, leather, industrial starch.

developed European countries, and reaches 50% or more amongst poorer households in more backward regions and countries. Despite scale economies which have led to the geographical concentration of food-chain activities[7], to the detriment of local food chains, the widespread nature of farming and of the consuming population means that this economic activity is important in all but the most urbanised locations of Europe.

Socially and culturally, agriculture occupies a unique role[8] as the main traditional "way of life", from which many community habits, structures and even language are derived. In most European countries, the majority of families have a member engaged currently or recently in farming, and the location of the leisure activities of millions is influenced by the availability or familiarity of agricultural land and buildings. These preferences are receiving a new interpretation in the light of growing environmental interest and concerns (see below), but underlie the popularity of rural tourism.

Environmentally, European agriculture is a land management activity carried out at varying latitudes and altitudes, and in both densely populated and more remote areas. It thus influences, in a huge variety of ways, the quality of natural resources such as land, water and air, the degree of biodiversity, and landscape characteristics. *"Agriculture remains a major source of pressure on the environment ... becoming even more intensive and specialized"* (EEA, 2001). Areas of general concern include:

- emissions of pollutants, particularly greenhouse gases, and fertiliser run-off into water systems
- lower population levels of both rare and once-common wildlife species, particularly birds as indicator species
- loss of "traditional" landscapes due either to simplification (e.g. removal of field boundaries, more monoculture) or to abandonment (desertification) or degradation (unused terracing, farm buildings, etc.).

However, the environmental role of European agriculture is positive as well as negative, with cropping systems (particularly traditional ones) maintaining specialised wildlife habitats, and in providing a basis for many residential and leisure activities. Moreover, these roles may be interpreted at various geographical levels, e.g. "macro" (part of the "European model of agriculture"), "meso" (part of national/regional characters) and "micro" (local environments).

Many of these developments were observable, though proceeding at different speeds and in different ways, in the ex-socialist countries before the start of transition in 1989. Since then, in the light of legal and economic uncertainties, agricultural adjustment in these countries has been patchy, but macroeconomic stabilisation and EU accession (including CAP adoption) seems likely to bring about similar trends in the accession countries.

---

[7] And farming itself, to some extent.

[8] In some regions, this role is occupied by fishing or hunting.

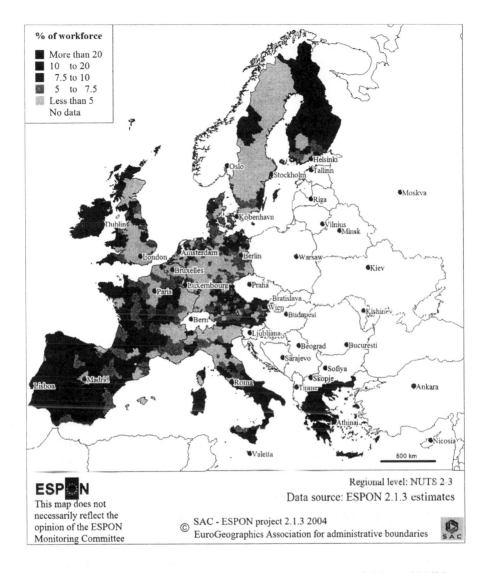

Map 2.1: Percentage employed in agriculture, forestry and fishing, 1995/96

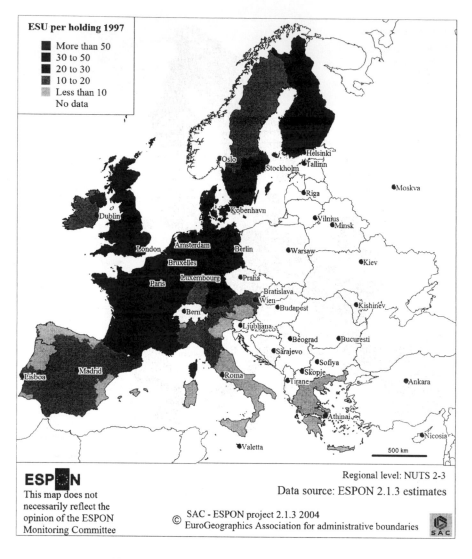

Map 2.2: Average size of holding in ESU, 1997

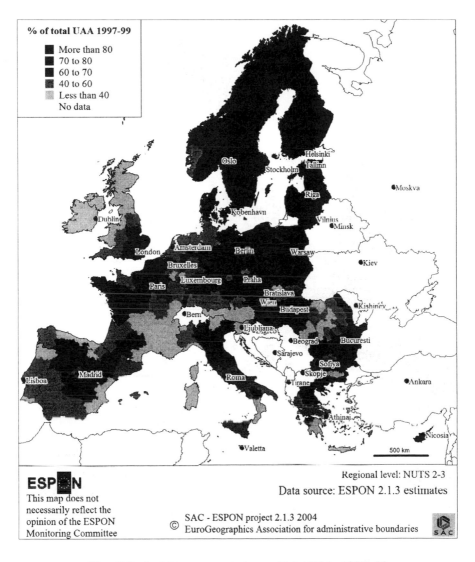

Map 2.3: Arable as a percentage of total UAA, 1997–99

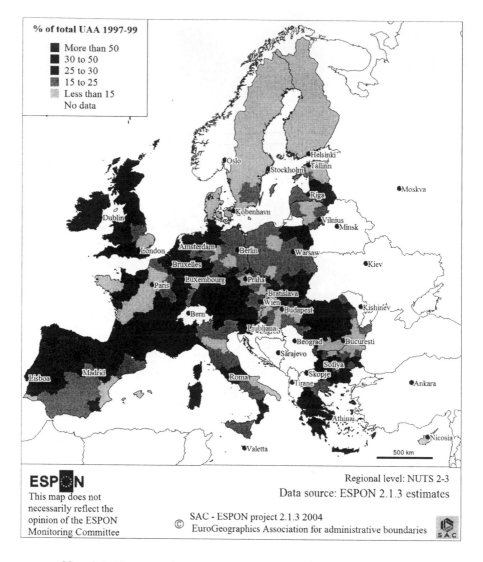

Map 2.4: Permanent grass as a percentage of total UAA, 1997–99

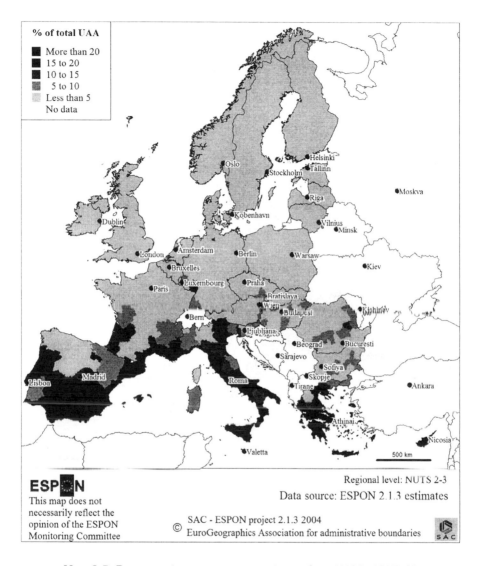

Map 2.5: Permanent crops as a percentage of total UAA, 1997–99

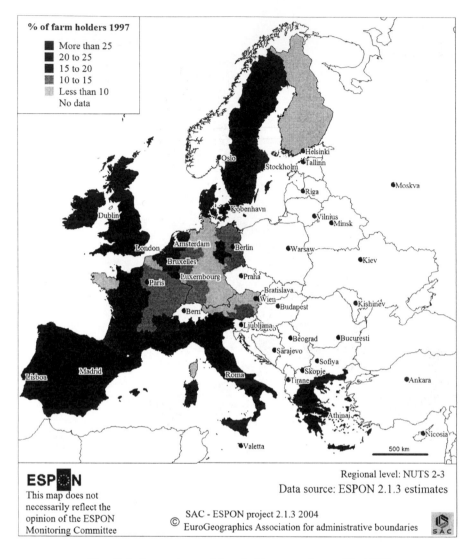

Map 2.6: Percentage of farmholders aged over 65, 1997

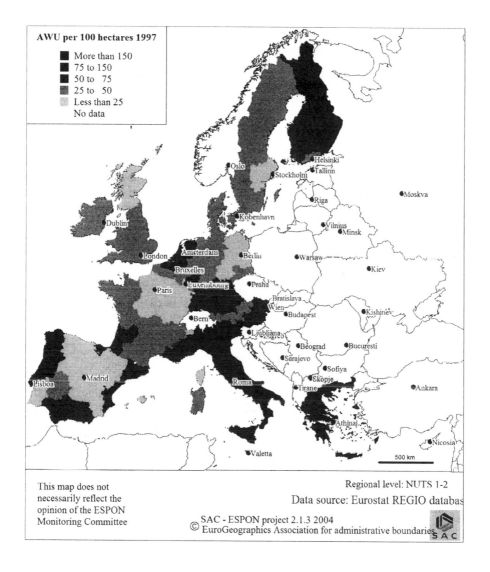

Map 2.7: AWU per 1000 hectares of UAA, 1997

2.2. The Common Agricultural Policy (CAP) and Rural Development Policy (RDP)

The scope of the EU's CAP/RDP for the purposes of this book is taken to be the interventions in farming undertaken by the European Commission for the purposes of pursuing Community objectives as set out in the various EU Treaties. These interventions can be categorised into three types:
a)   *expenditures from the European Agricultural Guidance and Guarantee Fund, EAGGF;*
b)   *market price support via non-expenditure methods such as tariffs and import quotas; and*
c)   *relevant EU Regulations and Directives.*

Given this definition, the following relationships between included and excluded policy areas are worth mention:

- Other structural policies: Regional and Social Funds are now partly "integrated" with EAGGF funding in Objective area "Programmes".
- LEADER: farming and farmers were involved to a greater or lesser extent in the previous LEADER and LEADER II Community Instruments (CIs) of the previous two budget periods (1988–93 and 1994–99); the current (2000–2006) LEADER+ scheme is funded entirely (except for national contributions) from within the EAGGF.
- EU environmental policy, e.g. the 1979 Birds Directive and the 2000 Water Framework Directive: environmental conservation and promotion (and sustainable development) are now over-arching EU policy objectives, and, in principle, all CAP initiatives must now carry environmental statements, and are subject to environmental criteria in their evaluation.[9] Agri-environmental CAP instruments, introduced as "accompanying" measures in the MacSharry reforms and expanded subsequently, have explicit environmental effects as their objective, and have attempted to regulate environmental threats posed by farming practices, and to encourage more environmentally friendly practices, such as organic farming, and low-intensity farming in areas of high nature value. However, with more "cross-compliance" (so far limited in uptake by Member States), this distinction between the two may erode in the future.
- EU competition policy: the Single Market is enforced with a set of regulations to control state (national and regional) aids; some such aids (which are inherently territorial) have persisted for special reasons. In the food chain, including farmer marketing agencies (e.g. the UK milk boards),

---

[9] However, a number of studies (e.g. Efstratoglou *et al.*, 1998; Elbe *et al.*, 2003) have shown that these regulations have so far had little effect.

the regulation of mergers and monopolies can fall under EU as well as national auspices.

• Food policy: there is increasing EU interest and active policy involvement in this area, largely through the Consumer Affairs DG. Regulations extend from farm (e.g. livestock welfare) through distribution and processing (livestock transport, slaughterhouse hygiene) to food retailing (e.g. traceability, labelling), including (e.g.) the regulation of organic food supply.

• National legislation: each Member State has its own set of laws regarding, for example, farm business taxation, land tenure/transfers and territorial planning regulations.

The original objectives of the CAP were laid down in Article 39 of the 1957 Treaty of Rome and in the conclusions of the 1958 conference at Stresa (Italy). The Article 39 objectives were (and are, since the Treaty remains in force, though subject to re-interpretation):

• increasing agricultural productivity
• ensuring a fair standard of living for farmers
• stabilising markets
• guaranteeing food security
• ensuring reasonable prices for consumers.

The Final Resolution at Stresa maintained that agriculture should be regarded as an integral part of the economy and as an essential factor in social life (Fennell, 1997, p. 20).

The underlying philosophy of the Common Market as a whole was to exploit comparative economic advantages, which include spatial differences in farming productivity in terms of soil quality, climate, distance from markets, etc. These factors clearly varied greatly from location to location within the original six Member States, and do so even more greatly within the EUs of 15 and 25. In pursuing the Treaty of Rome objectives, three "principles" were and still are commonly cited:

Market unity (or common pricing) involves the abolition of internal barriers to trade, and the establishment of common standards for food safety, quality, labelling, etc. In the first three decades of the CAP, as national currencies fluctuated against each other, complex agri-monetary measures and "green" exchange rates were introduced. However, with the achievement in 1992 of the Single Market, and the creation of the Euro as a single currency for 12 Member States in January 2002, these problems have largely subsided.

Community preference reflects the establishment of the European Community as a single customs union, with a common external tariff applied to all third-country imports as an instrument of market protection. Despite CAP reforms agreed since 1992, many rates of market protection are still high, particularly for sugar, milk, and beef. Nevertheless, trade preferences have been awarded to an increasing number of non-EU countries, some on historical grounds, some as part of pre-accession arrangements (e.g. with Bulgaria, Romania and Turkey), and some for reasons of economic assistance and development (e.g. the Maghreb of North-west Africa and ex-colonial African, Caribbean and Pacific countries).

Common funding (or financial solidarity) involves the use of income from the EU's "own resources" (mainly VAT- and GNP-based tax revenue, but also from import and other agricultural levies, some regarded as CAP "negative expenditure"), and expenditures via the European Agricultural Fund.

Producer co-responsibility has occasionally been promoted as a fourth CAP principle, i.e. that farmers should bear some of the financial burden of market support. Co-responsibility levies on marketings have long applied in the sugar regime, and for some years were operated in the dairy and cereal regimes. Nowadays, with the 2003 CAP reforms (see "Current CAP reforms" section below), it is more common to apply *cross-compliance* requirements, i.e. to be eligible for CAP payments, farmers must observe a range of management obligations, usually of an environmental nature.

None of these principles (except environmental cross-compliance) carry obvious territorial characteristics, and indeed they each imply an increased degree of common rather than differentiated treatment across the entire EU area, e.g. in terms of free production and trade in farm products. However, as the territorial implications of the CAP (and of other influences, such as technological change, and budget pressures) slowly became clear, the original "common" design of the CAP had to be adapted.

With the entry to the European Community of the United Kingdom along with Ireland and Denmark in 1973, a substantial area of "difficult" farmland, often with pre-existing policy measures in place, became subject to the CAP. Thus, in 1975, Directive 268 authorised the definition of certain agricultural regions as "mountainous" or *"less-favoured"* areas (LFAs), entitled to special direct payments to ensure "the continuation of farming". This marked the important departure – especially in the context of the present study – from the common policy treatment of farming in different parts of the Community. More details are given in the following section.

The 1987 Single European Act (Article 130R) mandated the consideration of environmental protection in all EU policies including the CAP/RDP. This led to the creation of a number of agri-environmental CAP measures (see below), and to a stronger (but still weak) element of environmental conditions in some other measures, e.g. stocking limits. These considerations led naturally to the specification of some new territorial aspects to the relevant CAP measures, mostly using the LFA boundaries.

In the Agenda 2000 reforms of the CAP, the "European Model of Agriculture" was endorsed, with objectives including:

- more market orientation and greater competitiveness
- food safety and quality
- stabilised agricultural incomes
- integration of environmental concerns into agricultural policy
- developing the vitality of rural areas
- simplification of administration, and

- strengthened decentralisation.

The Agenda 2000 reforms were followed, as scheduled, by the 2002/03 Mid-Term Review (MTR) of the CAP (CEC, 2002, 2003), in which the Commission argued that the objectives for EU agriculture should be:

- enhanced competitiveness
- more market orientation
- more sustainability
- a better balance of support, and
- strengthened rural development

to be achieved through:
- a single farm payment (SFP) independent from production,
- payments being linked to environmental, food safety, animal welfare, health and occupational safety standards,
- more money for rural development policy
- new measures promoting food quality, animal welfare and environmental standards,
- reduction in direct payments for bigger farmers, and
- further revisions to CAP market policy.

Both these sets of reforms are described in some detail in the next section. Here, it is simply noted that the new measures can be, and to some extent are, more spatially differentiated than has been possible (given the Single Market) with the traditional instruments of market support. The shift towards territorial considerations has also been promoted by principle of subsidiarity endorsed in the 1993 Amsterdam Treaty (Rabinowicz *et al.*, 2001), and national and regional governments have an increasing role to play in the preparation of various CAP/RDP arrangements for Commission approval.

The development of the CAP over recent years can be seen in the well-known Producer Support Estimates (PSEs) and Consumer Support Estimates (CSEs) prepared annually by the OECD. The PSE is the "monetary value of gross transfers from consumers and taxpayers to agricultural producers, measured at the farm-gate level, arising from policy measures that support agriculture, regardless of their nature, objectives or impacts on farm production or income" (OECD, 2002d). It is estimated commodity by commodity as the value of market price support (estimated from the differences between domestic prices and actual world prices) plus the sum of payments actually made to farmers on a variety of bases, e.g. output or input levels, or historical entitlement, and can be expressed as a percentage of gross farm receipts. The CSE is the corresponding measure on the consumer side (still measured at the farm-gate), i.e. the value of transfers to (or from) consumers of agricultural commodities.

Since the EU's percentage PSE hit a peak of 39% in 1986–88 (i.e. nearly 40% of the domestic value of EU farm output was represented by taxpayer and consumer support above external market price levels), the level of this measure has changed relatively little over the years, fluctuating in the 30–40% range. A provisional value of 37% has been reported for 2003 (OECD, 2004), somewhat above the OECD average of 31%. However, as alternative means of CAP support have been introduced, the share of market price support in total EU PSE has declined over time, from 87% in 1986–88 to 57% (of a total of €108 billion) in 2003. Over the same period, the CSE, which is largely based on price support, has fallen from 40% to 30%.

It should be noted in particular that EU budget expenditure on the CAP, i.e. the "taxpayer cost", is not an adequate measure of the "cost of CAP", since it excludes the major (and in fact dominant) PSE component arising from the effects of import barriers, quotas and other non-expenditure instruments in raising domestic EU prices for farm products above the levels obtaining outside the EU.

2.3. CAP/RDP measures and expenditures

Agenda 2000 defined two "pillars" of the CAP. Pillar 1 comprises:

- commodity market support regimes with intervention buying or private storage aids
- "lightweight" regimes with emergency buying and producer group support
- direct payments, often with quotas and/or reference yields and area ceilings to limit expenditure
- supply management tools such as quotas on milk supplies, maximum stocking densities and compulsory arable set-aside
- other elements such as environmental or animal welfare requirements, "outgoer" (e.g. dairy) schemes and grubbing-up aid.

Pillar 2 covers measures under the Rural Development Regulations (RDRs) 1257-1260/1999, such as:

- aids for farming in Less-Favoured Areas and now in areas with environmental restrictions
- agri-environment schemes
- support for farm forestry
- aid for farm investment, modernisation, and diversification
- aids for marketing and processing
- early retirement aids, and aids for young farmers
- vocational training
- aids for improved water management, land reparcelling and land improvement (Article 33 of Regulation 1257/1999)
- support for developing farm-related tourism and craft activities (Article 33)
- other farm-related rural development provisions (Article 33)

The "common rules" Regulation 1259/1999 authorises "modulation" to switch funding from Pillar 1 to certain elements of Pillar 2 (Article 4). However, modulation was initially implemented only by France – which later suspended the process – and by the United Kingdom. Since the reforms agreed in June 2003, modulation is to become mandatory for all countries, operating on the new Single Farm Payment (SFP) by means of a flat-rate cut rising from 3% to 5% in 2007. Governments may supplement this by additional national modulation.

Table 2.1 shows expenditures from the Guarantee Section of the EAGGF by Member State for 2001, classified by commodity and other sector. The main item, accounting for over 40% of the total of €42 billion, relates to "arable crops", i.e. cereals, oilseeds and protein crops (peas and beans), and is mainly direct area-based payments (including those on set-aside), with a relatively small amount of market support expenditure on export refunds and storage. The next highest item relates to bovine meats, i.e. beef and veal (mainly direct payments), and smaller commodity-related expenditures to olive oil (mainly direct payments), milk products (market support), fruit and vegetables (market support), sugar (market support), sheep and goat meat (mainly direct payments), wine (market support) and tobacco (mainly direct payments). EAGGF Guarantee expenditure on rural development measures (previously accompanying measures) accounts for about 10% of the total.

For the purposes of this project, it is often convenient to distinguish CAP/RDP measures into six categories, on the basis that each has potentially different territorial impacts:

- market regulation: the "traditional" CAP instruments of market support for most (but not all) farm commodities via import taxes, export subsidies and intervention purchasing, together with secondary measures such as marketing quotas. The major economic effect is not via subsidy expenditure, but via higher internal prices maintained by border measures; these are regularly estimated by the OECD's PSEs.
- direct income payments: made directly (or nearly so, e.g. to cooperatives, etc.) to farmers linked to production, e.g. area and headage payments. Various constraints, such as set-aside for commercial arable farmers, and stocking densities for grazing livestock payments, are attached to these payments. Under Agenda 2000, these payments may be "modulated", i.e. reduced for individual farmers in order to finance Pillar 2 activities, but this has not yet been widely undertaken.
- LFA payments: a dual-purpose instrument, addressing both environmental and socio-economic goals
- agri-environmental schemes: "accompanying measures" introduced originally at the time of the 1992 CAP reforms, and currently paid under the Rural Development Regulation (RDR) 1257/1999
- rural development measures[10], including other "accompanying measures" (early retirement and afforestation) as well as those for farm development

---

[10] In current Commission parlance, the term "rural development" is used very widely, to include both agri-environmental land management and "real" development in rural areas, whether on-farm or off-farm (e.g. diversification). It has also been used to encompass food quality and animal welfare, which are likely to become of increasing importance. In the context of this study, however, "rural development" was taken to

and diversification, food processing and marketing, training, the broader
Article 33 measures for village renewal etc., and LEADER
- other measures, e.g. input subsidies and (farm-specific) taxes.

The LFA and the three accompanying measures mentioned above account
for about 50% of the funding of Rural Development Programmes in all EU
countries. However, the situation in the member countries differs substantially;
the Netherlands have the lowest share (13%) and Ireland more than 90%. All such
measures are part-financed (in differing proportions from country to country, and
region to region) by the EU, the rest being made up of national-government (and
private) funds.

Figure 2.1 shows how funds currently flow from the two Sections of the
EAGGF to the various measures of the CAP/RDP. It should be noted that Pillar 2
is currently funded by both the Guarantee and Guidance Sections, which have
very different budgetary and administrative arrangements, e.g. co-financing.
Also, the large component of expenditure devoted to direct aids (€27.4 bn) is
being increasingly decoupled from specific commodities and their output levels,
and is conditional on increasing environmental regulation.

---

cover measures and payments for processing and marketing; training and diversification;
farm development; Article 33; and LEADER.

Table 2.1: CAP expenditures by Member State, 2001 (million Euro)

	B	DK	D	EL	E	F	IRL	I	L	NL	A	P	FIN	S	UK	Total*
Arable Crops	166	666	3739	483	1934	5181	120	1919	11	251	379	242	353	420	1603	17466
Sugar	281	86	237	8	62	357	4	143		50	28	21	10	23	187	1497
Olive Oil				587	1030	5		848				54				2524
Dried Fodder etc.		10	23	5	186	83		48		14		1			4	375
Textile Plants	9		2	543	212	42				4	1	3			9	826
Fruit, Vegetables	37	1	17	235	522	294	2	348		40	2	42		2	17	1558
Wine Products			41	16	470	222		378			14	54			1	1197
Tobacco	3		34	376	115	77		339			1	19		9	9	973
Other Crop Products	3	32	18	24	52	26		118		10		5	2	2	4	297
Crop Products	499	794	4111	2276	4584	6287	126	4144	11	368	425	441	365	448	1824	26713
Milk and Milk Products	181	128	186	–3	29	500	144	92		479	–27	–3	46	28	127	1907
Bovine Meat	169	83	744	61	734	1468	827	296	8	86	172	126	62	101	1116	6054
Sheep and Goat Meat	1	1	34	201	390	144	90	143		12	4	48	1	3	374	1447
Pig Meat, Eggs, Poultry	5	26	5	1	11	52	1	8		19	4	3		1	2	137
Fish					6	3	1					1				13
Livestock Products	356	238	969	261	1171	2167	1063	539	8	597	153	175	109	134	1619	9559
Non-Annex 1 Products	40	33	65	3	23	53	51	19		79	19	2	6	9	36	436

Table 2.1 (cont)

	B	DK	D	EL	E	F	IRL	I	L	NL	A	P	FIN	S	UK	Total*
Food Programmes	8	2	17	15	63	65	2	49		2	1	28	7	9	12	282
Ultra-Periphery Progs.				24	90	39		1				30				184
Vet. & Phytosanitory	4	3	22	4	18	27	15	24		51	2	8	1	1	383	566
Fraud Control & Prvtn.			10	3	11	–1	–1			1	2	–1			10	32
Reductions in Advances	–2		–27	–45	–311	–40		–143		1						–570
Promotion, Information	1		5		4	5	1	1		4	1				3	49
Other Measures	1	8				39	17	57					1	29	318	470
Total Other	51	46	91	4	–103	184	84	7	1	136	24	69	15	48	753	1448
Rural Development	32	35	708	75	540	609	327	660	10	55	453	197	327	151	184	4363
Total* FEOGA Guarantee	938	1114	5880	2616	6194	9248	1599	5349	30	1155	1055	882	816	780	4381	42083

Source: 31st Financial Report on the EAAGF Guarantee Section, 2001 Financial Year, Annexe 8, COM(2002)594.
* Individual values may not add exactly to Totals, due to rounding and/or small amounts unallocated to countries.

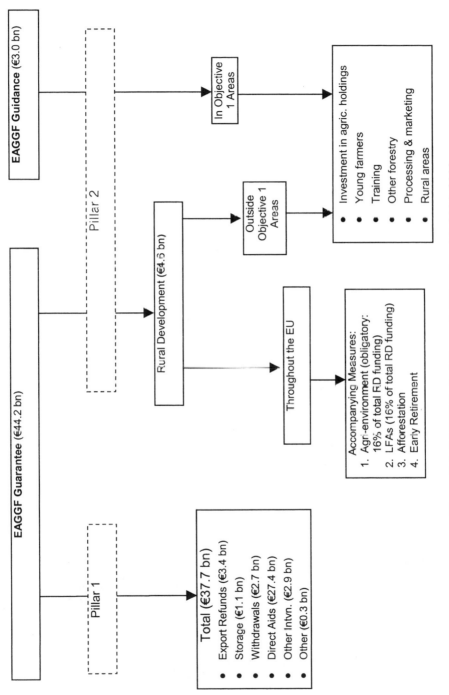

Figure 2.1: EAGGF Budget Flows via Pillars 1 and 2, 2004–2006

EAGGF Guidance (€3.0 bn)

EAGGF Guarantee (€44.2 bn)

Pillar 2

Pillar 1

In Objective 1 Areas

Rural Development (€4.6 bn)

Outside Objective 1 Areas

Throughout the EU

- Investment in agric. holdings
- Young farmers
- Training
- Other forestry
- Processing & marketing
- Rural areas

Accompanying Measures:
1. Agri-environment (obligatory: 16% of total RD funding)
2. LFAs (16% of total RD funding)
3. Afforestation
4. Early Retirement

Total (€37.7 bn)
- Export Refunds (€3.4 bn)
- Storage (€1.1 bn)
- Withdrawals (€2.7 bn)
- Direct Aids (€27.4 bn)
- Other Intvn. (€2.9 bn)
- Other (€0.3 bn)

   The shifting of policy priorities towards Pillar 2 of the CAP, as addressed above, is an on-going process, with major differences between Member States and regions. The Rural Development Regulation has thus become an increasingly important component of the CAP. Differences between countries are due not only to different policy priorities, but also to their varying resource bases and experiences with the development and application of rural development measures and programmes. The RDR is sometimes seen as a tool to promote environmental land management, whereas others focus more strongly on the modernisation of agriculture. Planned expenditure on Pillar 2 has been analysed by Dwyer *et al.* (2002) and indicates considerable differences between Member States in the allocation of RDR spending to the various objectives, as shown in Figure 2.2. Member States had to draw up Rural Development Plans and Programmes[11] at the appropriate geographical level, which was broadly interpreted as either national or regional. The regional option was taken up by the UK, Italy, Spain, Germany and Belgium.

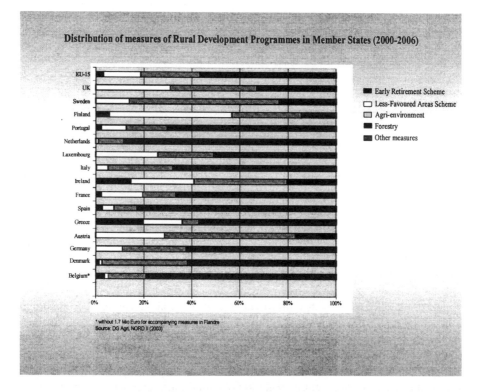

Source: Dwyer *et al.*, 2002, p.19.

Figure 2.2: Distribution of RDP Measures in Member States, 2000–2006

---

[11] Both these are often represented by the acronym RDP, which in this book is reserved for the EU's Rural Development Policy.

The differences in the regional priorities of rural development measures are particularly articulated in Germany, where the variation of allocation between the Länder regions shows similarities to the picture at EU level. Poorer Länder prioritise Article 33 measures (notably village renewal schemes) while wealthier ones emphasise agri-environment and LFAs (Dwyer *et al.*, 2002, p. 22). A similar calculation on preliminary data for application of the RDR in the New Member States (2004–2006) also revealed variation in policy implementation.

2.4. Territorial components of CAP/RDP measures

Any CAP measure may have differential effects over the Community space, depending on the presence and nature of agricultural activity. By definition, market support in the Single European Market, without intra-EU border controls and measures such as the previous "green" exchange rates, are largely non-territorial, except insofar as some of these instruments, which operate at EU borders and at intervention purchasing points, may favour EU producers near these locations due to transport costs.

Nevertheless, as indicated above, several CAP/RDP measures have strong territorial characteristics, in being applicable, at different rates, or at all, in various parts of the Community. In some cases (e.g. sugar quotas), the spatial element is restricted to Member State level, with complete freedom of action within national borders; in others, such as Less-Favoured Areas (LFAs) or Objective 1 areas, there are more detailed geographical specifications.

National and regional ("ring-fenced") quotas for milk and sugar have obvious territorial characteristics, being based on historical levels of production in the various areas defined in the regulations. In some countries, the growth of a relatively free market in such quotas will have minimised the territorial "quota effect" when compared to the spatial pattern which would have emerged without quotas (but with price support); in others, the lack of such a market will have enhanced it by "freezing" production patterns down to farm level. Similar effects can be expected with eligibility "quotas" for farm grazing livestock numbers, and with some "maximum guarantee quantities" (tobacco, etc.).

The current arable regime includes regionally specified "reference" crop yields as the basis for rates of direct payments, and hence has a territorial character, though one that offsets regional agronomic differences that would otherwise have meant a "biased" application of the direct payment system. The impact of this feature will depend on the "coarseness" of the regions defined by Member States when this regime was introduced, and possibly the interpretation for the purposes of policy implementation of "good farming practice" criteria.

The first initiative to introduce an explicitly territorial dimension into the CAP was the Council Directive 75/268/EEC on Less-Favoured Areas. Prior to the mid-1970s, the regional dimension of European agricultural policy, though recognised and indeed emphasised by the operation of common (i.e. horizontal) price and structural instruments available everywhere within the Community of

Six, had not been reflected in the CAP. The ill-fated "Mansholt Plan" of 1968 had envisaged some regional differentiation, and the modernisation Directive of 1972 contained the possibility (taken up only by Ireland and Italy in 1974) of operating higher rates of reimbursement in problem areas. However, according to Fennell (1997, p. 253), "the practical response to the needs of the backward regions was completely inadequate".

The entry of the United Kingdom, along with Ireland and Denmark, to the Community in 1973 forced a change in this approach. With the aim of maximising domestic production and thus national self-sufficiency in food, the UK had operated a hill-farming policy, in the shape of headage payments and additional capital grants, ever since the Second World War, and would have been most unwilling to give this up. However, in the accession negotiations, the UK was careful to argue that any new CAP measures of this kind would apply to areas in other countries with similar problems. Consequently, as early as February 1973, the Commission proposed a draft Directive on farming in mountain areas and in certain other poorer farming areas. Later, it widened the scope of the measure from securing farm incomes to continued conservation of the countryside and to maintaining the population in disadvantaged regions. After discussion over the methods to be employed (annual compensatory payments based on livestock units, and special investment aids), on rates to be paid, and on the areas to be covered (both left to Member States to propose within Community guidelines), Directive 75/268 on mountain and hill farming and farming in *certain less-favoured areas* was passed. As explained above, this marked a major change in the nature of the CAP by introducing regional categories. It also represented the initiation of direct annual payments to farmers, an approach which was to expand greatly in the 1990s and thereafter. However, unlike the later direct payments, LFA rates could be altered from year to year, to take account of fluctuating conditions. The LFA designation was also used to effect various territorial modifications to other CAP instruments (e.g. modernisation grants, and later co-responsibility levies) by granting more favourable treatment of LFA farms.

Three types of LFA were established: mountain areas where "erosion" and "leisure needs" were specified as protection objectives, areas in danger of depopulation, and "other smaller areas" affected by specific handicaps and where farming was needed to preserve tourist potential or to protect the coastline. The Guidance Section of the Community's Agriculture Fund (FEOGA) reimbursed 25% of the total cost (35% in Italy and Ireland; and for a number of regions, particularly in Objective 1 areas and other regions of Cohesion countries, a substantially higher co-financing was later agreed).

As a complement to the range of sectoral support measures already in place, the LFA Directive provided a framework for payment of annual compensatory allowances to farmers in less-favoured areas. Specifically, Directive 75/268 stated that:

> "...the steady decline in agricultural incomes in these areas as
> compared to other regions of the Community, and the particularly

poor working conditions prevalent in such areas are causing large-scale depopulation of farming and rural areas, which will eventually lead to the abandonment of land that was previously maintained.... The permanent natural handicaps existing in such areas, which are due chiefly to the poor quality of the soil, the degree of slope of the land and the short growing season and which can be overcome only by operations the cost of which would be exorbitant, lead to high production costs and prevent farming from achieving a level of income similar to that enjoyed by farms of comparable type in other regions... It may be essential, if the objectives assigned to farming in the less favoured areas are to be attained, that farmers permanently engaged in agriculture in such areas be paid annual compensatory allowances".

Directive 75/268 was incorporated into the subsequent Regulation 797/85, and, following the reform of the Structural Funds (see Chapter 3) in 1988, the LFA scheme was incorporated as part of a horizontal EU Objective 5a measure under the Structural Funds. In 1991, Regulation EEC *no.* 2328/91 provided for extra payments in designated LFAs characterised by one or more of the following attributes:

1.   Permanent handicaps (altitude, poor soils, climate, steep slopes),
2.   Undergoing depopulation or having very low densities of settlement, and
3.   Experiencing poor drainage, having inadequate infrastructures, or needing support for rural tourism, crafts and other supplementary activities.

After modification by Regulation 950/97, LFA legislation was consolidated under Agenda 2000 into Articles 13–20 of Regulation 1257/99. The main change was the move from headage to area-based payments in order to cut off the link with production and to avoid incentives to raise production.

Over the years, the area designated as LFA has expanded, partly due to the accession of further Member States with a particularly high share of LFAs, but also because states have proposed extensions to their LFAs. By the early 1990s, about 55% of the Community of Twelve was so classified, a portion which has remained equal since then (Dax and Hellegers, 2000, p.182). Map 2.8 shows these areas, expressed as a proportion of the total area in each NUTS3 region in the EU-15. Within the 10 New Member States, the relevance of LFA support is even higher than for the EU-15: whereas up to now about 19% of EU-15 support through the rural development programmes is spent on the LFA instruments on average, the New Member States have planned to concentrate more than 25% of rural development support for LFA support.

The objectives of the LFA Compensatory Allowances are "to maintain a viable agricultural community and thus help develop the social fabric of rural areas by ensuring a fair standard of living for farmers and by off-setting the effects of natural handicaps in mountain and less-favoured areas".

In 1999, the total expenditure on Objective 5a throughout the EU-15 was €1310.9m, or 23.5% of the total EAGGF Guidance Section expenditure. For the period 2000–2006, when spending was switched to the Guarantee Fund, there is provision to allocate €924m (3.8% of the total for rural development policy) to LFAs and areas with environmental restrictions in Objective 1 regions, and €4631.9m (18.9%) to Non Objective 1 regions.

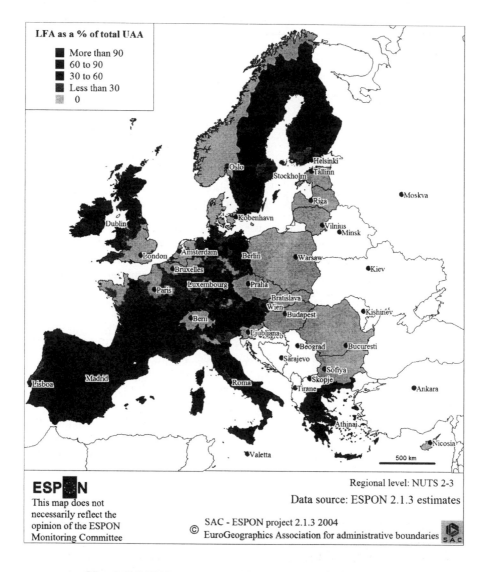

Map 2.8: LFA Area as a % of Total Area by NUTS3 Region

2.5. Current CAP reforms

The Council of Agricultural Ministers agreed a substantive reform to the CAP in June 2003. These reforms are examined in detail in Chapter 6 but are briefly summarised here. Further cuts in support prices for crops and livestock will be accompanied by the replacement of the multitude of various direct payments by a single partially decoupled direct payment (the Single Farm Payment, or SFP). Compulsory modulation of all SFPs will rise to 5% by 2007, and the funds saved will be spent under Pillar 2 rural development programmes. In order to reflect different national circumstances, Member States were given considerable discretion in how these reforms are implemented. For example, the SFPs can be related to historical levels of payment to each farm, or calculated as a flat-rate per hectare, or a hybrid of these two.

Table 2.2 shows the position proposed to the Commission by each EU-15 Member State at time of writing (late 2004) as regards their implementation of the options permitted by the 2003 CAP reform decisions. The majority have decided to adopt the "historical" basis for SFPs, i.e. maintaining individual farmers' receipts during 2000–2002. Others have chosen a "static hybrid" model, which combines a historical payment component with a "flat-rate" per-hectare component (usually on a regional basis), or a "dynamic hybrid" model which moves towards the flat-rate over several years. Moreover, most Member States (the exceptions are Ireland and the UK other than Scotland) have also chosen to maintain some degree of "coupling" support payments to products (crop areas or livestock numbers). Some Member States have also opted to use a "national envelope" to "ring-fence" coupled funds for particular livestock practices.

It will be seen that the main feature of these proposals (with which the Commission is expected to agree, in general) is to maintain the general geographical pattern of CAP support, especially at the meso and macro scales. At the micro scale, the adoption of SFPs means that some farms (e.g. horticultural enterprises), which hitherto receive no or low payments, will receive significantly more. Also, the implementation of a direct payment system for dairy cows, which is to be incorporated into the SFP system, will somewhat alter the overall geographical pattern of CAP payments, though not necessarily that of total CAP support, since the new dairy payments are meant to compensate for lower market price support.

Table 2.2: Summary of 2003 CAP reform decisions

Country (regions)	Start	SFP Basis	Coupling Rates, Notes
Belgium	2005	historical	100% suckler cows, 100% calf slaughter, 100% seeds (partial)
Denmark	2005	static hybrid	75% special male cattle, 50% sheep
Germany (Länder)	2005	dynamic hybrid, to FR	25% hops, 60% tobacco (until 2009)

Table 2.2 (cont)

Country (regions)	Start	SFP Basis	Coupling Rates, Notes
Greece	2006	*historical*	*40% durum wheat, 50% sheep*
Spain	2006	historical	25% arable crops, *100% seeds,* 100% suckler cows, 100% calf slaughter, 40% adult cattle slaughter, 100% for all products in outermost regions
France	*2006*	*historical*	25% arable crops, 50% sheep, 100% suckler cows, 100% calf slaughter, 40% adult cattle slaughter. 100% for all products in overseas territories
Ireland	2005	historical	no coupling
Italy	2005	historical	100% seeds, NEs for arable crops (7%), beef (8%) and sheep
Luxembourg	2005	static hybrid	no coupling
Netherlands	*2006*	*historical*	*100% calf slaughtering, 40% adult cattle slaughter, 100% seed for linseed*
Austria	2005	historical	100% suckler cows, 40% adult cattle slaughter, 100% calf slaughter, 25% hops
Portugal	2005	historical	100% suckler cows, 40% adult cattle slaughter, 100% calf slaughter, 50% sheep, 100% seeds, 100% for all products in outermost regions
Finland (3 regions)	2006	*dynamic hybrid*	75% special male cattle, *10% arable crops?, 100% seed?, 50% sheep? 10% NE for quality beef*
Sweden (5 regions)	2005	static hybrid	74.55% for special male cattle, 0.45% NE for beef
United Kingdom			
- England (3 regions)	2005	dynamic hybrid, to FR	no coupling
- Scotland	2005	historical	10% NE for quality beef
- Wales	2005	historical	no coupling
- N. Ireland	2005	static hybrid	no coupling

Notes:
1.   Entries in *italics* indicate informal notification only to Commission. Question marks (?) were in the originals
2.   SFP = single farm payment; NE = national envelope; FR = flat-rate (area)
Source: Agra Europe, 2117 (6 August 2004) p. EP/7.

2.6. Pre-accession and enlargement aspects of the CAP/RDP

The accession of ten new Member States (eight Central and Eastern European Countries, NMSs, plus Malta and Cyprus) to the EU took place in May 2004, with Bulgaria and Romania possibly acceding in 2007. Prior to accession, these countries have been preparing their agricultural sectors and policies for EU entry and CAP adoption, e.g. by instituting CAP-like support systems, and seeking liberalised trade with the EU-15. Each started with its own national structure of agriculture and agricultural policy, in the NMSs often with significant differences between conditions in the early 1990s and those in the mid-2000s. The territorial aspects of agricultural and rural development policies in the accession states are therefore complex.

The current policies of the CAP seem hardly suitable for the structural problems of the NMSs. The discussion in the negotiation period has concentrated on the application of CAP in the accession countries and transition periods useful for the sector and spatial integration. Rural development policy attains particular relevance under these circumstances, since it is assumed that a great portion of regions in the NMSs will be affected by further spatial divergence tendencies.

Community policies on agricultural enlargement, as defined by Agenda 2000, focus almost entirely on the combined capacity of overall growth in the NMSs and the structural aid to relax these constraints, by absorbing agricultural over-employment in urban and rural (non-agricultural) employment, and by financing increased national budgets for agricultural modernisation and restructuring (Pouliquen, 2001, p. 83).

Transitional policies for the accession countries are proposed to achieve a competitive restructuring process of the sector, and to focus on measures in favour of rural development and government aid for the transformation of the semi-subsistence-farming sector in order to keep the migration towards urban employment on a moderate level. The competitive restructuring covers direct aid to investment in intensive productions, notably livestock and horticulture, and in the related upstream and downstream industries. Basic infrastructures (networks of water conveyance, electricity, roads, rail and waterways, telecommunications, irrigation, and other para-agricultural investments) are of particular importance and furthermore a complex "package" of other convergent policies, including relevant progress of the institutional framework are required (Pouliquen, 2001, p. 83).

Given the expenditure and non-expenditure effects of the CAP, the main effect of EU accession and CAP adoption in the new Member States derives from Pillar 1, i.e. market policy and direct payments (which are being made on a simplified basis in most of the new countries). As regards Pillar 2, NMS applicants prepared for EU entry by setting up regional authorities for the development of rural development programmes, used to implement pre-accession funding via the Special Action for Pre-Accession measures for Agriculture and

Rural Development (SAPARD) programme agreed at the European Council meeting in Berlin as part of the Agenda 2000 proposals. In addition, a Special Preparatory Programme (SPP) in the framework of PHARE has been established (in the years 1998 and 1999), which among other things financed capacity building, training and technical assistance for the preparation of a national Rural Development Plan in each applicant country. This plan served as a basis for measures under the SAPARD programme.

The SAPARD programme disposed of about €520m per year, and acted through horizontal measures towards the adaptation of agricultural structures and policy as well as support for rural development. In most applicant countries, the required national co-financing (25%) for both funds took up a large part of the current budgetary resources for these measures (about 14% of national agricultural budgets in all the NMSs; Dwyer *et al.,* 2002, p.100). SAPARD is analysed in Chapter 5.

As for the EU-15, the development of the rural population and economy of a specific region in the new Member States is strongly linked with overall employment opportunities in these regions. Quite diverse situations and trends between regions occur in these countries, and there is rising concern about an increase in regional differentials between agglomeration and marginal areas which is even greater than in the EU-15. With regard to cohesion objectives, this poses considerable challenges for application of EU policies.

Despite the awareness of the increasing problems of rural areas in these countries the main focus of national policies has been laid in the transformation period on economic growth indicators, the adaptation of structures, the improvement of administrative capacities and processes and compliance with EU standards. This has resulted in an on-going trend towards differentiation of growth regions and many rural, more peripheral regions lagging behind in regional development. Some recent regional classification studies have confirmed this experience and pointed to the increasing spatial divergence of new Member States (Baum *et al.*, 2004; Bika, 2004; IRDSP, 2004).

The need for a stronger commitment to the problems of rural regions was realised as a major factor for the preparation of the countries to EU accession. Preparation for membership of the EU required substantial changes to industrial and public infrastructure, administrative institutions and procedures, as well as training and capacity-building programmes. To support these often costly and time-consuming measures, the EU has established PHARE, which has become a familiar source of funding. Two further funds (SAPARD and ISPA) were agreed at the European Council meeting in Berlin as part of the Agenda 2000 proposals. In addition a Special Preparatory Programme (SPP) in the framework of PHARE has been established, which among other things financed capacity building, training and technical assistance for the preparation of a national Rural Development Plan in each applicant country. This plan served as a basis for measures under the SAPARD programme. These programmes were meant to have a great influence on the spatial development policies of the new Member States. However, delays in the negotiation process of the respective programmes and the focus on improving market structures resulted in a limited impact on rural economy. As no minimum requirements were set for the use of more innovative measures these were only applied to a rather restricted extent.

The current NMS Rural Development Plans (2004–2006) provide the opportunity to implement the whole range of rural development measures. There is, however, a continued concentration on restructuring and competitiveness measures, with a rising concern for environment and land management measures. For many countries, the competitiveness of the agricultural sector is still the major concern, and rural development issues are gaining in importance only gradually. The rural economy measures (mainly for diversification activities, and renovation and development of villages) are of minor concern in these programmes, and measures such as LEADER+ are only foreseen in some of the countries. However, the preparation process for such long-term, bottom-up processes has started in some NMSs, e.g. Hungary, Czech Republic.

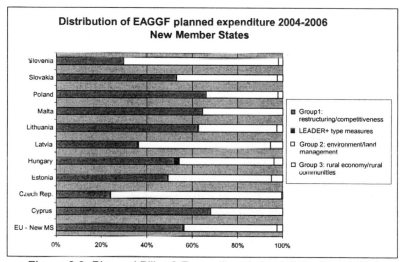

Figure 2.3: Planned Pillar 2 Expenditure in the NMSs, 2004–2006

The considerable potential of rural amenities in the new Member States has been addressed by research projects and country studies reflecting the concern to preserve the natural resource base and regional economy of rural regions. These include strategies for the preservation of high nature value farming systems and areas of high environmental value in great parts of the rural regions. From a long-term perspective, it seems of particular relevance to address these features as a major potential of the rural regions. Spatial considerations will thus be of primary importance for the development of the next Rural Development Programmes for the period 2007–2013. These will have to refer to the Commission's actual proposals for future rural development policy (see Chapter 6). As these set requirements for minimum shares of the national envelope to be spent on each of the three rural development axes (and also for the LEADER element), more balanced strategies will have to be found.

Chapter 3

3. EU Structural Policies and the ESDP

One of the fundamental objectives of the EU is to construct a competitive economy, based on democracy and the rule of law, which will benefit its entire people. In the words of Article 2 of the 1957 Treaty of Rome, it is intended:

> to promote throughout the Community a harmonious development of economic activities, a continuous and balanced expansion, an increase in stability, an accelerated raising of the standard of living and closer relations between the States belonging to it.

Various strategies, such as the Single Market (abolishing barriers to trade between Member States), European Monetary Union (e.g. a single currency), and the funding of research and technology programmes, are intended to promote the above objective by means of centralised management.

However, economic theory and reality both suggest that economic progress is unlikely to proceed evenly throughout the territory of the EU, and may even widen gaps between the richest and the poorest groups, whether defined regionally or socially. Indeed, lack of comparative advantage, and inability to adjust quickly to changing domestic and external pressures, have meant that development which has bettered some producers or consumers in the EU has disadvantaged others, leaving them less prosperous in relative or even absolute terms. Even before the recent enlargement, the per capita gross domestic product (GDP) of Luxembourg was twice that of Greece. Currently, income levels in the more prosperous countries of the EU-15 are over twice those in several New Member States, even allowing for "purchasing parity".

To counter these trends, and in particular to assist those "lagging behind" the EU income average, the EU has instigated a number of structural funds and policies, designed to improve the capacity of regions and social groups to compete effectively within the Union by assisting in the mobilisation of resources (natural, technical and human) and otherwise developing their economic processes. These instruments are not explicitly designed to be "compensatory" in nature, i.e. they seek to "invest" in new capabilities which will eventually be able to survive without assistance, rather than acting as welfare supports or tax concessions (which are still the prerogative of Member States).

Nevertheless, the identification and funding of special groups and regions from the central EU budget is bound to produce intensely political debate, often expressed in nationalistic terms. Several approaches have been taken in attempts to handle these difficulties, such as co-funding (Member States have to produce a share of the required expenditure), formulae or criteria based on "common" criteria (e.g. unemployment rates), the use of sectoral funds (for agriculture and fisheries), and the Cohesion Fund, which assists environment and transport projects in the 13 Member States whose GDP is below 90% of the Community average[12].

Even so, the extent to which the richer Member States are prepared to contribute to funds from which they will receive little direct benefit themselves is responsible, at time of writing, for a major struggle over the size of the EU budget for the 2007–2013 period, i.e. whether it should be planned on the basis of contributions amounting to 1.14% of national GDP levels (as proposed by the Commission), or only 1% (as proposed by the United Kingdom, Sweden and some other countries), i.e. over 10% less, at a time when poor countries have become members. Despite considerable recent economic success, some older Member States such as Spain and Ireland feel that the 2004 enlargement has endangered their net receipts of structural funds, as poorer countries such as Poland present their stronger claims.

Whatever the outcome, debate will continue on how available funds should be allocated, spent and monitored. Discussion and analysis is not helped by the long-term nature of the "multi-annual programmes" used to disburse structural funds in an effort to avoid "stop-start" effects, and indeed the inherent difficulty of discerning the effects of only one policy component of the wider economic environment determining economic progress in regions "lagging behind". Nevertheless, as the structural funds begin to challenge the CAP FEOGA Guarantee Section in their shares of the overall EU budget (see below), it is important to appreciate their nature and significance for the present study.

3.1. Policies and funds

The Structural Policies of the EU are aimed at reducing disparities between different regions and social groups, and at promoting sustainable development and general economic efficiency. They are financed from the Regional Development Fund (originated in 1975), the Social Fund and the Guidance Section of the Agriculture Fund (both established since the origins of the European Community), together with the Cohesion Fund and the Financial Fund for Fisheries Guidance (FIFG) (established in 1993). There are also two pre-accession financial instruments, designed in 1999 for the 12 countries then applying for accession: the Instrument for Structural Policies for Pre-Accession (ISPA) and the Special Action for Pre-Accession measures for Agriculture and Rural Development (SAPARD). There are also some smaller Community Instruments (CIs) funded directly from the Commission rather than through Member State governments, with similar objectives.

---

[12] Czech Republic, Estonia, Cyprus, Latvia, Lithuania, Greece, Hungary, Malta, Poland, Portugal, Slovenia, Slovakia, Spain.

These funds are intended:

- to reduce disparities in development and promote economic and social cohesion in the European Union.
- to improve the effectiveness of the Community's structural assistance by concentrating the assistance, and simplifying its allocation by reducing the number of priority Objectives.
- to identify more precisely the responsibilities of the Member States and the Community at each stage: programming, monitoring, evaluation and control.
- During the 1994–99 period, the three main Structural Funds concentrated on a number of key "objectives", namely:
- Objective 1 – structural adjustment and development of less developed regions
- Objective 2 – conversion of regions severely affected by industrial decline
- Objective 3 – combating long-term unemployment and facilitating the occupational integration of young people and persons excluded from the labour market
- Objective 4 – assistance for workers in employment to adapt to industrial change and new production systems through retraining
- Objective 5a – speeding up the adjustment of agricultural and fisheries structures
- Objective 5b – facilitating development of rural areas, and
- Objective 6 – promotion of development in regions with exceptionally low population density.

Of these seven Objectives, nos. 1, 2, 5b and 6 were spatially restricted in their remit, while no explicit spatial restriction applied to Objectives 3, 4 and 5a (the so-called "horizontal" measures).

Objective 5a was explicitly targeted at the agricultural industry, and continued long-standing funding of capital investment on and off farms by means of grants and loans. Rural areas with economic problems fell either into Objective 1, where more integrated development programming was attempted, or, with lower rates of support, into Objective 5b. In addition, the LEADER II Community Initiative (CI) focused on "bottom-up" projects in rural areas. The Cohesion Fund was focused on environmental and transport projects in countries with GDP per head under 90% of the EU average, i.e. Greece, Portugal and Spain.

For the period 2000–2006, the objectives were rationalised down to three:

- Objective 1 – programmes in regions whose development is lagging behind, including regions whose per capita GDP falls below 75% of the EU average, sparsely populated regions of Finland and Sweden and the most remote regions,
- Objective 2 – programmes in regions undergoing conversion including industrial or service sectors subject to restructuring, a decline in traditional activities in rural areas, problem urban areas, and difficulties in the fisheries sector,

- Objective 3 – modernising training systems and promoting employment outside the regions eligible for Objective 1.

while the LEADER + Community Initiative continues to promote rural development through the initiatives of local action groups (LAGs).

The total appropriation for the Structural Funds and the Cohesion Fund for the period 2000–2006 stands at €213 billion at 1999 prices, i.e. approximately €30 billion per year, of which about €2.5 billion is allocated to the Cohesion Fund. Almost 70% of the main component goes to Objective 1, and about 12% to each of the other two Objectives, with the remainder going to Community Initiatives such as LEADER (about € 300m per year) and to FIFG outside Objective 1 areas.

3.2. Structural funds in rural areas

In the two previous programming periods, 1988–93 and 1994–99, structural funds were allocated in rural areas via Objectives 1, 5a and 5b, and 6, although some of the Structural Fund expenditure under the other Objectives (2, 3 and 4) had indirect effects on rural areas. LFA expenditures were in fact the major EAGGF Guidance Section commitments, and approximately €15 billion were allocated to rural development over the 1988–93 period. Between the 1989–93 and 1994–99 periods, Objective 5b areas were considerably expanded (approximately doubling in area and population, overall, taking into account the accession of three new Member States in 1995). On the accession of Sweden and Finland in 1995, Objective 6 was added for Nordic areas characterised by extremely low population densities (no more than 8 persons per square kilometre).

In Objective 1 areas, rural development measures were financed, within an integrated (i.e. territorial) approach, from the EAGGF Guidance Section, with the exception of the Less-Favoured Area scheme for which the EAGGF Guarantee Section was used.

Table 3.1 shows commitment appropriations for the four CIs for the programming period 2000–2006, by Member State. It can be seen that the main rural CI, LEADER, accounts for about 20% of total CI appropriations. An additional but unknown share of INTERREG funding to cross-border, transnational and interregional cooperation purposes will also be applied in rural areas.

Table 3.1: Indicative allocations of commitment appropriations among the Member States, 2000–2006 (in million euro – 1999 prices)

	LEADER	INTERREG	EQUAL	URBAN	TOTAL
B	15	104	70	20	209
DK	16	31	28	5	80
DE	247	737	484	140	1608

Table 3.1 (cont)

	LEADER	INTERREG	EQUAL	URBAN	TOTAL
GR	172	568	98	24	862
E	467	900	485	106	1958
F	252	397	301	96	1046
IRL	45	84	32	5	166
I	267	426	371	108	1172
L	2	7	4	0	13
NL	78	349	196	28	651
A	71	183	96	8	358
P	152	394	107	18	671
FIN	52	129	68	5	254
S	38	154	81	5	278
UK	106	362	376	117	961
Networks	40	50	50	15	155
EUR-15	2020	4875	2847	700	10442

Source: European Commission Press Notice, Brussels, 13 October 1999, no. IP/99/744.

3.3. The European Spatial Development Perspective (ESDP)

According to Faludi and Waterhout (2002), the European Spatial Development Perspective (ESDP) "towards balanced and sustainable development of the territory of the European Union" is the culmination of "years of dedicated work" by spatial planners in Europe since the 1970s and 1980s. Initiated by the French and Dutch with their centralist heritage, the Germans and others were brought in for a succession of annual meetings starting with those at Nantes and Turin in 1989 and 1990. At Leipzig in 1994, the "Principles" of the ESDP were agreed, establishing balanced spatial development as a key to economic and social cohesion. "Spheres of activity" included "a more balanced and polycentric urban system", "parity of access to infrastructure and knowledge" and "wise management ... of Europe's natural and cultural heritage". The British, who had hitherto been lukewarm or hostile to the ESDP concept, began to play a more positive role after 1997. And the ESDP itself was finally agreed by an informal Council of Ministers meeting at Potsdam as "a suitable policy framework for the sectoral policies of the Community and the Member States that have spatial impacts, as well as for regional and local authorities" (EC, 1999).

There have been frequent disputes over the competencies of EU institutions to embrace the planning of land use, transport, housing and other elements of spatial planning. The principle of subsidiarity forces ESDP proponents to argue the case for extending EU powers (i.e. national and regional government constraints) into these areas. So far, the ESDP has not been formally embedded into an EU Treaty; hence the "informal" nature of the Potsdam meeting, and the lack of specific ESDP Directives, let alone a dedicated Directorate-General in the Commission.

A second reason for the hesitant and patchy acceptance of the ESDP is its "very ambiguous geographical imagination" and "contradictory discourses" (Zonneveld, 2000; Jensen and Richardson, 2000, cited in Healey, 2004). There appear to be major problems of vocabulary and language translation (and so perhaps understanding) surrounding terms such as "spatial planning", "strategic", and "city", let alone more complex concepts such as "polycentricity" and "territory". Some of these problems may arise from dissonance between the concepts of traditional (or "essentialist") geography of place and those of the newer relational geography that focuses on links and networks of flows (Healey, 2004). Others derive from the ongoing discussion within geographical economics as regards location, agriculture, and rural development (Kilkenny, 2004).

A third constraint on effective action on spatial planning at EU level derives from lack of adequate data to cover the multi-dimensional concerns of planners. Depending on the context, and perhaps viewpoint, planners are both developers in the indirect sense of helping to determine the delivery of state-funded infrastructure, and state-empowered regulators of private-sector development, trying to ensure the provision of public goods and services (or to avoid public "bads" and disservices). The EU-wide NUTS (Nomenclature des Unités Territoiriales Statistiques) system still has only a limited and inconsistent relationship to national and regional systems of data collection, e.g. administrative, political, agricultural, and environmental. At a simpler level, even if DG Agri and Eurostat have developed over many years some harmonised statistical systems of agricultural census data, commodity market supply-demand balances and farm business accounts, this has not yet extended to all other sectors. Thus, for example, there is no EU-wide database of Natura 2000 sites, and many alternative definitions of "distance" (accessibility, remoteness).

Many of these problems have of course worsened considerably with the entry of the ten New Member States (NMSs) in May 2004. Even if data exists, there is so far only limited progress in collecting these into a common database usable at either the "macro" or EU scale, or, for comparative purposes, at lower "meso" (national/regional) or "micro" (regional/local) scales.

3.4. ESDP concepts applied to agriculture and rural development

The above brief historical review of the ESDP has involved a number of general terms and concepts, which, for the purposes of this study, must be applied, from a territorial point of view, to agriculture and rural development, and to the relevant policies. Such a discussion may be either theoretical or empirical; the paragraphs below seek to combine these two treatments.

Competitiveness: from an economic point of view, the ability to earn income (profits, increased wealth) from the resources of a particular area depends partly on the market demand for the various products and services which may be supplied, and partly on the efficiency with which the region's resources can be utilised, including the important dynamic aspects of marketing and product innovation.

The well-known Engel's law of economics recognises that, with growing incomes, demand for food (especially at farm-gate level) grows less than proportionately. This feature of European and worldwide markets has resulted in stagnant demand for many farm commodities, and hence in reduced competitiveness of rural regions as regards food production, whose prices in real terms have fallen almost relentlessly for many years. Demands for alternative land-using output, such as wood, fibre and biomass, have not compensated for reductions in the value of raw materials for food and drink.

European territories dependent on agriculture have thus experienced a long-term decline in competitiveness as regards their traditional products. Moreover, the effects have not been equally distributed in spatial terms: marginal (in geographical terms) regions have tended to suffer most, as a result of more difficult natural conditions (and particularly slack demand for red meats from grazing livestock). The establishment of the Single European Market has allowed the forces of comparative advantage within agriculture to favour those better placed geographically (climate, transport) and in terms of policy support (milk, cereals) while disadvantaging others (mountainous and some Mediterranean regions).

It is difficult to see that spatial planning, however exercised, can prevent these long-term developments. The CAP, at great budget cost, and with a number of territorially specific instruments introduced to offset the more brutal effects of interregional competitiveness, has not prevented a steady drain of labour from the agricultural industry. Thus relative incomes for the remaining farming population have remained reasonably stable in real terms. More direct "planning" interventions might hinder over-intensive exploitation of certain regions (e.g. by strict landscape and water regulations), and assist others more disadvantaged (e.g. by poor communications), but are unlikely to be more successful.

Instead, the concept of competitiveness as it affects rural areas in Europe applies more and more to "new" and non-commodity economic activity, in particular residence away from the congestion, pollution and high prices of the conurbations, and leisure, either short-term (day trips) or long-term (tourism). This "consumption of the countryside", using income derived elsewhere ("export" services, public sector salaries, pensions, dividends, etc.), alongside a certain amount of small-scale manufacturing, has the potential to revive the economies of many (but not all) rural areas. In this respect, a "spatial planning perspective" has much to recommend it, in order to discipline the operation of uncontrolled market forces which might lead to both the over-exploitation of certain areas through over-development and the under-exploitation of others through lack of infrastructure.

Territorial Cohesion: As described in the initial section of this chapter, agriculture occupies a small but central role in the economic, social and environmental character of countries and regions, and is thus a determining factor in achieving (or not) cohesion – "the more balanced distribution of activities" – between territories. At the same time, the highly varied nature of agriculture (in comparison to, say, modes of urban transport, or household living patterns), means that allowance must be made for the inevitable (and desirable) heterogeneity of farming from place to place.

From an economic point of view, the above discussion on competitiveness suggests that overall territorial cohesion will only be attained by accepting, and adjusting to, the reducing role of agriculture as an economic industry in many areas. A spatial planning perspective must therefore identify alternative uses of land, buildings and people (human resources) within "territories", and help to design region-specific plans, regulations and fiscal systems working towards the encouragement of necessary adjustment.

Countrywide (national) planning is likely to be much too crude to be able to achieve this level of detail; much smaller regions seem more appropriate[13]. One difficulty at the "meso" and even "micro" levels is that "natural" agricultural regions (and hence the environments they create), such as "upland" and "lowland", or "peri-urban" or "distant", tend not to overlap conveniently with most planning and administrative boundaries[14]. Thus, territorial cohesion has to be interpreted flexibly as regards the agricultural sector.

Polycentricity (the "promotion of complementary and interdependent networks of towns as alternatives to the large metropolises or capital cities, and of small and medium-sized towns which can help integrate the countryside" (DG Regio, 2004): As can be deduced from the quoted definition, this ESDP concept arose from concerns about over-urbanisation, primarily at the 'macro' European scale in which the "Blue Banana" of central EU cities threatened to leave other EU cities lagging behind in economic and other terms. Rural areas, and the agriculture within these, play little or no role at this level. The reference to "small and medium-sized towns which can integrate the countryside" appears to imply that such settlements can act effectively as routes by which rural areas and villages can be assisted at a "micro" level; but this still leaves agriculture, and the countryside generally, as essentially residual in nature.

From an agricultural and rural perspective, both urban-rural relations generally and the role of settlements in the countryside have long been a focus of economic and sociological research. The von Thünen model of a central place supplied largely from its own hinterland may have become outdated with cheaper

---

[13] It is notable that, in many EU countries, such as France, Germany and the United Kingdom, regional agencies and assemblies have been given more powers over agricultural and rural development in recent years.

[14] Thus Germany has introduced a regional competition "*Regionen Activ*" on sustainable integrated rural development which not only allows but encourages the self-definition of problem or development regions independent of administrative boundaries, even between states (Länder).

transport, but the Lösch/Christaller model of a hierarchy of settlements with differing levels of service provision still holds sway. More recent thinking concerns "key settlements" as "growth poles", possibly in "networks" or "partnerships", and the revival of "market towns" as a preferred settlement type.

From an agricultural point of view, it is somewhat difficult to promote polycentrism as an obviously more efficient and desirable objective for food production: modern methods of farm production and long-distance transport have rendered the nearby proximity of settlements to farms largely redundant. Socially, the loss of the farming population, as mentioned above, has reduced the sense and utility of a "farming community", but this does not relate directly to "centres". A much stronger preference for local foods, perhaps based on concerns over food safety and quality, or on (much) higher fuel costs, might suggest that more and smaller settlements could benefit farms more widely, but the prospect seems unlikely.

Much more promising are the more recent trends in residential and leisure patterns, which favour rural space within easy reach (and preferably within view). Driven largely by higher-income social groups seeking privacy, security, quiet and recreation, rural development can become a widespread phenomenon, with or without "polycentres", although services such as schools and supermarkets may (but need not) lead to thriving settlements.

As argued above, the role of spatial planning in this context must be (amongst others) to prevent "congestion" in the general sense of the term, i.e. a reduction in the average quality of life due to the unfettered actions of individuals seeking private optima, and to ensure the provision of the necessary infrastructure (hard and soft) insofar as this cannot be efficiently provided through the marketplace for lack of a pricing mechanism or appropriate institutions.

Chapter 7 returns explicitly to the ESDP concepts when reviewing the results of the analysis.

Chapter 4

4. The Territorial Distribution of CAP/RDP Support

4.1. Introduction

This chapter considers the way in which CAP support has, in practice, been distributed across European territory, where farmers operate in a wide range of economic, social and environmental contexts. In addition to differing farm input and output prices (despite the Single Market), there are considerable differences in natural production conditions, and high variation in agricultural structures and production methods. Thus, analysis was expected to highlight territorial imbalances in the incidence of the CAP and RDP support. However, the extent and nature of these imbalances were difficult to predict. The results presented in this chapter are based on a statistical analysis of indicators and data at NUTS3 level over the period 1990 to 2000, augmented by findings from an EU-wide review of literature.

Since the different types of support (market price support, direct income payments, agri-environmental payments etc.) have each played a distinct role within the CAP reform process and may have given rise to territorially distinct effects, the chapter considers the incidence of Pillar 1 and Pillar 2 support separately before considering more generally the influence of farm type, accessibility and region type on the distribution of support. The final section of the chapter summarises the findings.

4.2. Data

4.2.1   Data sources, coverage and apportionment

The availability of detailed territorial data on agriculture across Europe is surprisingly poor, given the extent of agricultural data collection and the bureaucratic burden on farmers. Very little data relating to agriculture are available at NUTS3 level from Eurostat, DG Regio or DG Agriculture, and where they do exist up to 91% of data are missing. DG Agriculture reported that they have no information on CAP expenditure below national level other than Farm Accountancy Data Network data, which shows support received rather than expenditure. As a result, the process of compiling an appropriate dataset takes considerable time and much effort. Importantly, it required drawing on national and OECD sources and the use of apportionment methods (described in the next section).

Many of the data sources have incompatible geographies. For instance, both the FADN and the EUROFARM use hybrids of NUTS1/2/3. In the case of the EUROFARM database, these are known as "Districts".

As noted above, very little of the raw data in the REGIO database is available at NUTS3 level. Indeed, the only indicator from this dataset widely available at NUTS3 level relating to agriculture is employment in agriculture, forestry and fishing (derived from the Regional Accounts). Similarly, the FADN dataset only provides data at NUTS2 or NUTS1 level, and sometimes in non-standard areas. Finally, data on CAP and RDP expenditure is not available at NUTS3 level.

Data from EUROFARM dataset, containing results from the EU surveys of the structure of agricultural holdings, provides a far richer source of indicators on the agricultural sector at NUTS3 level. However, the EUROFARM dataset relates only to "old" Member States, not N12 or EFTA countries and, even in relation to the EU-15, is incomplete. Therefore, it was necessary to develop a method to apportion the indicators available on the REGIO, FADN or CAP/RDP databases from NUTS1 or NUTS2 level to NUTS3 level.

The method chosen for apportionment of higher-level data on farm numbers, crop areas, livestock numbers, subsidy receipts, etc. to NUTS3 level was based on the following core set of agricultural land-use variables available at NUTS3 level from either EUROFARM or national sources:

- arable area (ha)
- permanent crop areas (ha)
- utilised agricultural area (ha)
- number of dairy cows
- total number of beef animals (or total cattle less number of dairy cows)
- total number of sheep and goats
- number of agricultural holdings/farms
- number of agricultural work units (or agricultural employment).

The actual variable used to allocate an indicator from NUTS2 to NUTS3 depended on the indicator to be apportioned. For example, in the case of disaggregating the total level of feed used for grazing livestock, the sum of beef and sheep numbers was used, on the assumption that the relative proportion of total grazing livestock is consistent with the relative proportion of feed used in each component NUTS3 region. Similarly, in allocating total cereal compensation payments from NUTS2 to NUTS3 level, the arable area of each NUTS3 region was used as the apportionment variable. As indicated by these examples, the method relies on the assumption that the actions of farmers (in relation to feed per livestock unit in the first case, and enrolment on the arable payments scheme in the second) do not vary significantly within each NUTS2 region (or, alternatively, vary to the same extent within each NUTS3 region).

4.3. Territorial typologies

Several types of territorial typologies are available for use in grouping NUTS3 regions on the basis of their demographic, economic and agricultural data. Some of these are in widespread and/or official use, others require specific analyses to be carried out. A crude version related to the CAP would have been to group each NUTS3 region according to whether or not it fell into an LFA region, in whole, in part, or not at all. Another would have been to carry out cluster analyses of the NUTS3 data, using all available information (see Shucksmith *et al.*, 2004). The European Commission has a scheme based on a population density threshold of 100 persons per sq. km. However, for this study, it was decided to use a combination of a well-known typology established by the OECD with an economic dimension, and an urban-rural typology developed by other ESPON research programme partners. These are described below in turn.

4.3.1   The OECD typology

The OECD scheme covers the entire territory of a country, not just the "rural" part, and distinguishes two levels of geographic detail. In a first stage, each basic administrative or statistical unit, in most cases the commune or community, is first classified into "rural" and "urban" by the simple criterion of a population density threshold of 150 inhabitants per sq. km. In a second stage, the degree of rurality of a region comprising several or many of such relatively small communities is classified by the share of the population living in rural communities, thus distinguishing the following three types of regions:

- predominantly rural regions (more than 50% of the population living in rural communities),
- significantly rural regions (between 15% and 50% of the population living in rural communities) and
- predominantly urbanised regions (less than 15% of the population living in rural communities).

This distinction between the hierarchical levels of territorial detail is central to the conceptual approach of the territorial typology. Only through the different levels can the complexity of rural problems in various national and regional contexts be seized. The framework is conceived also to allow for analysis of interrelationships between regions but to enable differentiation between rural and urban communities within a region at a lower geographic level.

There was also explicit recognition that "territorial development performance is not strictly correlated with the degree of rurality or urbanisation" (OECD, 1996a, p. 53).

In a final step of analysis, population and employment changes were thus chosen to serve as primary indicators to offer an indication of the prospects for regional development and lead to a further differentiation into leading and lagging regions within each of the previous types. The simple split was done by comparing the regional performances with the respective national averages, which led to six types of regions (at NUTS3 level):

- Predominantly Rural + Leading
- Predominantly Rural + Lagging
- Intermediate + Leading
- Intermediate + Lagging
- Predominantly Urban + Leading
- Predominantly Urban + Lagging

4.3.2   An urban-rural area typology

The urban-rural typology developed by another ESPON project (CURS, 2004) indicates an overall physical and functional pattern of Europe. Two sets of criteria were used in order to distinguish between different types of regions: the degree of human intervention and the degree of urban integration. There are six different types of regions:

- urban-rural, high urban integration
- urban-rural, low urban integration
- rural-urban, high urban integration
- rural-urban, low urban integration
- residual-urban, high urban integration
- residual-rural, low urban integration

As seen above, the variations in the physical environment (land use) are measured by the degree of human intervention. This is considered to be "high" in areas where the share of artificial surface is above the European average, "intermediate" in areas where the share of agricultural land is above European average and "low" in areas where the share of natural areas is above European average.

The degree of urban integration is measured by two different criteria: population density above/below the average (107 inhabitants/km^2) and the position of the most significant centre in the urban hierarchy (at least a European-level functional urban area; Nordregio, 2003b). The assumption is that the rank of an urban centre corresponds with its influence on the NUTS3 area. High urban integration would thus be characteristic to all NUTS3 areas with at least a European-level Functional Urban Area. The areas with population density above the European average have also been included in the category of high urban integration.

4.4. Pillar 1 support

4.4.1   Pillar 1 support in the EU-15

An initial hypothesis was that the distribution of the Pillar 1 support is not consistent with the economic or social cohesion objectives of the EU. To test this proposition, the relationship between the level of Pillar 1 support received by each NUTS3 region and GDP per inhabitant, unemployment rates and population change of each region was investigated. In this and subsequent analyses, Pillar 1 support is defined as the sum of the value of market price support (MPS) and direct income payments received by farmers.

Using the method described above, MPS data were derived from the apportionment of OECD EU-level data to NUTS3 regions, while the value of direct payments was derived from the FADN database apportioned to NUTS3 regions. To allow for differences in the scale of NUTS3 regions, the level of support per agricultural work unit (AWU) and per hectare of utilisable agricultural area (UAA) were taken as the basic units of analysis. The year investigated was 1999, prior to the Agenda 2000 reforms but well into the period following the MacSharry reforms of the CAP. The population change indicator for each NUTS3 region relates to the 1989 to 1999 period and has been weighted by the level of population in each region to allow for scale effects.

Table 4.1 reports the correlation coefficients between total Pillar 1 support and indicators of economic and social cohesion (GPD per head, unemployment rates and population change).

Table 4.1: Pearson correlation coefficients between level of total Pillar 1 support accruing to NUTS3 regions and socio-economic indicators, 1999

	GDP per head	Unemploy-ment rate	Population change 1989–99	Support per ha UAA	Support per AWU
Support per ha UAA	0.088(**)	−0.305(**)	0.216(**)	1	0.261(**)
N	1051	945	892	1051	1051
Support per AWU	−0.143(**)	−0.095(**)	0.117(**)	0.261(**)	1
N	1053	947	892	1051	1053

** Correlation statistically significant at the 0.01 level (2-tailed).

Concentrating first on the level of support per hectare UAA, the results indicate that, in 1999, total Pillar 1 support was distributed in such a way that it tended to benefit richer regions, regions with lower unemployment rates and regions with growing populations. They thus support the hypothesis that the incidence of Pillar 1 support acts in such a way that it does not contribute towards the economic and social cohesion objectives of the EU.

However, when considering support per annual work unit (AWU) employed in agriculture, the findings are somewhat different. In particular, whilst higher levels of support still seem to be associated with regions with lower unemployment rates and higher population gains, a significant negative correlation coefficient was found between support levels per AWU and per capita GDP. In other words, while Pillar 1 support per hectare goes unambiguously to richer regions, support per worker is distributed more ambiguously.

To explain this, consider the final two columns of Table 4.1, which show that support per AWU and support per ha UAA are not closely correlated with one another, due to substantial differences in the land and labour intensity of different agricultural production systems. This explains why different perspectives on the distribution of support are gained depending on which denominator is used in the analysis.

Map 4.1, showing Pillar 1 support per AWU, shows a concentration of support in northern areas of Europe, while the distribution appears more dispersed when expressed per ha UAA (Map 4.2). In the latter case, areas of northern Spain, parts of Italy and Greece are among the highest beneficiaries. In both cases, significant differences in the level of support received by farmers within national boundaries can be detected.

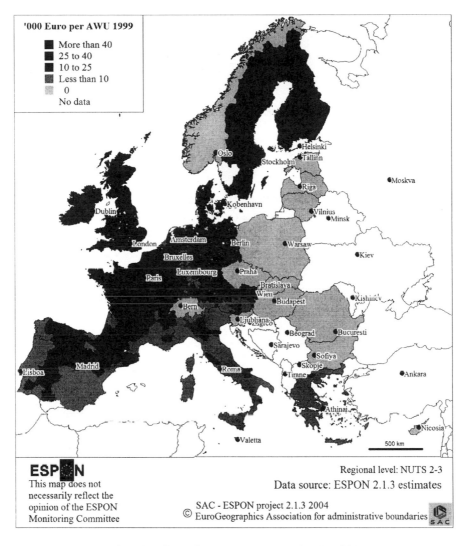

'000 Euro per AWU 1999
- More than 40
- 25 to 40
- 10 to 25
- Less than 10
- 0
- No data

ESP⬤N

This map does not
necessarily reflect the
opinion of the ESPON
Monitoring Committee

Regional level: NUTS 2-3
Data source: ESPON 2.1.3 estimates

SAC - ESPON project 2.1.3 2004
© EuroGeographics Association for administrative boundaries

Map 4.1: Total Pillar 1 support per AWU, 1999

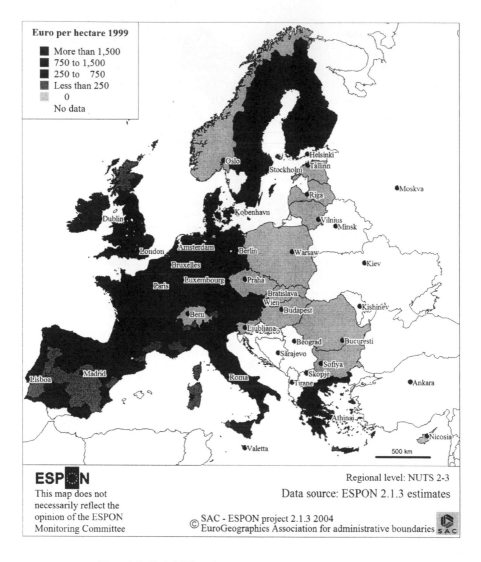

Map 4.2: Total Pillar 1 support per hectare UAA, 1999

In general, Table 4.1 confirms the expectation that Pillar 1 of the CAP, with its strong production focus, is not helping to achieve the cohesion objectives of the EU. While this may have been acceptable in the early decades of the CAP when policy objectives were focused on ensuring food security and economic efficiency, it is now, arguably, increasingly problematic. The agrarian concept of rurality that underpinned the CAP in the 1960s and 1970s is becoming less and less appropriate in the emerging context of a "post-industrial" European rurality (Sotte, 2003) and the limitations of the traditional CAP are especially problematic in the context of EU enlargement (Buckwell *et al.*, 1995).

4.4.2   Differences between Pillar 1 measures

Correlation analysis was used to investigate the relationship between the two policy instruments that comprise Pillar 1 of the CAP – MPS and direct income payments – and socio-economic indicators. The results, which are summarised in Table 4.2, indicate that the largest element of the CAP, market price support, like total Pillar 1 support, was distributed in a manner inconsistent with social and economic cohesion objectives.

Table 4.2: Pearson correlation coefficients between the level of market price support and direct income payments accruing to NUTS3 regions and socio-economic indicators, 1999

	Support per ha UAA	Support per AWU	GDP per head	Unemploy -ment rate	Population change 1989-99
Market Price Support					
Support per ha UAA	1	0.557(**)	0.113(**)	−0.371(**)	0.199(**)
N	1069	1068	1069	963	896
Support per AWU	0.557(**)	1	−0.089(**)	−0.161(**)	0.116(**)
N	1068	1078	1078	972	900
Direct Income Payments					
Support per ha UAA	1	0.382(**)	−0.156(**)	0.209(**)	−0.028
N	1067	1065	1067	961	894
Support per AWU	0.382(**)	1	−0.191(**)	0.163(**)	−0.103(**)
N	1065	1065	1065	959	893

** Correlation statistically significant at the 0.01 level (2-tailed).

However, direct income payments tended to be higher in areas with a low GDP per capita, with high unemployment rates and falling populations. Thus, direct payments were distributed in a manner which supports cohesion objectives. This supports the argument that the introduction of direct payments led to a more equitable distribution of support between regions of Europe by weakening the link between the level of aid to regions and their agricultural performance (EC, 2001c). However, direct income payments remain problematic for other reasons. Buckwell (1996) argued that the levels of payments have not been sufficiently linked to the income reductions associated with the lowering of commodity price supports. This has led in some circumstances to overcompensation of some groups of farmers, as acknowledged by the Commission in Agenda 2000. Further, he argues that there is insufficient rationale to support an indefinite continuation of such payments for a once-off policy change.

Map 4.3, 4.4 and 4.5 all confirm the distinct territorial incidence of Pillar 1 support instruments, showing separately Market Price Support, crop-related direct income payments and livestock-related payments, all expressed per AWU.

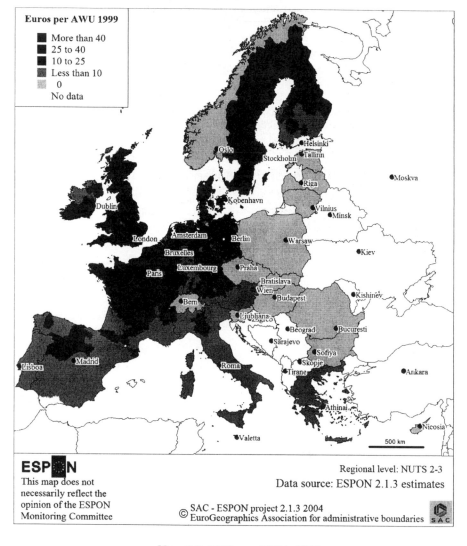

Map 4.3: MPS per AWU, 1999

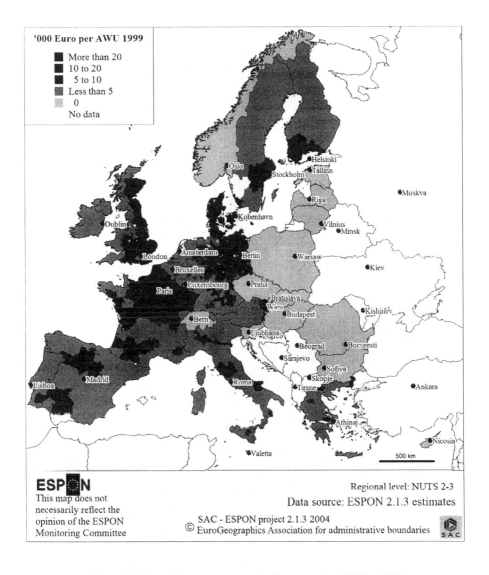

Map 4.4: Direct income payments for crops by AWU, 1999

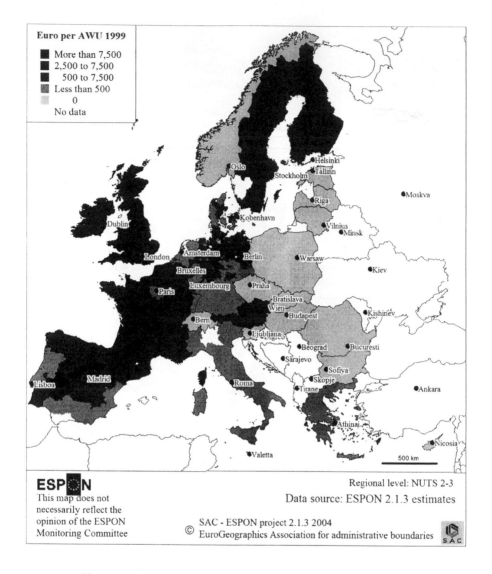

Map 4.5: Direct income payments for livestock by AWU, 1999

4.4.3   Pillar 1 support in the New Member States

Most accession states have been preparing their agricultural sectors and policies for EU entry and CAP adoption by instituting CAP-like support systems, and seeking liberalised trade with the EU-15. The territorial aspects of agricultural and rural development policies in the New Member States (NMSs) are therefore complex, with significant differences between conditions in the early 1990s shortly after the start of transition to those expected in (say) the mid-2000s.

Policy data at NUTS3 level relating to the NMSs is more scarce than that available for the EU-15. However, some preliminary regression analysis was carried out to test whether the national agricultural policies of some NMSs in 1999 were consistent with economic and social cohesion.

Table 4.3: Pearson correlation coefficients between socio-economic indicators and the level of market price support accruing to NUTS3 regions, selected New Member States

	Support per ha UAA	Support per AWU	GDP per head	Unemploy-ment rate
Czech Republic				
Support per ha UAA (14)	1	0.480	−0.722**	0.213
Support per AWU (14)	0.480	1	−0.688	0.106
Hungary				
Support per ha UAA (20)	1	n.a.	0.396	−0.329
Support per AWU (-)	n.a.	n.a.	n.a.	n.a.
Poland				
Support per ha UAA (44)	1	n.a.	0.083	−0.334*
Support per AWU (-)	n.a.	n.a.	n.a.	n.a.
Slovakia				
Support per ha UAA (8)	1	0.909**	−0.592	0.296
Support per AWU (8)	0.909**	1	−0.589	0.563
All				
Support per ha UAA (86)	1	0.659**	0.126	−0.106
Support per AWU (22)	0.659**	1	−0.380	−0.375

** Correlation is significant at the 0.01 level (2-tailed).
* Correlation is significant at the 0.05 level (2-tailed).

The results in Table 4.3 show clear differences in the way MPS was distributed across regions of the NMSs. In the Czech Republic, support in 1999 tended to be higher in areas with a low GDP per capita and with high unemployment rates (although the latter estimate is not statistically significant). In contrast, in Poland, support was higher in areas with lower unemployment rates. No statistically significant results were found in relation to the distribution of MPS in Slovakia and Hungary. When considered overall, MPS payments in the four NMSs in 1999 tended to be higher in areas with a high GDP per capita and with low unemployment rates. In other words, as in the case of the EU-15, the policy of MPS did not support objectives of improving economic and social cohesion. Further discussion of the impact of agricultural and rural development policy in the NMSs is given in Chapter 6.

4.5. Pillar 2 support

Since the CAP was not originally designed as a cohesion instrument, it could be argued that the above findings relating to Pillar 1 are not surprising. In comparison, Pillar 2, often hailed as representing a fundamental departure towards a more integrated rural development policy, might be expected to be distributed more in line with cohesion objectives. To test this expectation, the relationship between the level of Pillar 2 support received by each NUTS3 region and GDP per head, unemployment rates and population change of each region was investigated.

The level of Pillar 2 support was estimated in two ways: firstly, by the sum of the value of environmental subsidies and LFA payments *received* by farmers, again derived from the FADN database apportioned down to NUTS3 level, and secondly, through the apportionment of national Rural Development *expenditure*, taken from Dwyer *et al.*, 2002, in this case using FADN data as a means of distributing the country-level totals between regions. Neither approach to estimating Pillar 2 support is entirely satisfactory: the first because it is based on sample data and only takes into account LFA and agri-environmental payments, excluding other Pillar 2 schemes; the second because it is based on budget figures rather than actual expenditures. However, using both measures together provides a robust basis for understanding the territorial impact of Pillar 2 policies. The methodologies used for the apportionment of these data to NUTS3 regions are described in Section 4.2.1.

Correlation analysis, summarised in Table 4.4, shows that, contrary to expectations, Pillar 2 support as represented by the FADN-derived payments to farmers is inconsistent with cohesion objectives, favouring the more economically viable and growing areas of the EU. Pillar 2 support based on apportioned Rural Development budget data appears to be more equitably distributed, with a significant negative correlation coefficient between support per AWU and GDP per head, but again a negative relationship between levels of support and unemployment rates is observed: higher levels of support are associated with lower unemployment rates.

Table 4.4: Pearson correlation coefficients between total Pillar 2 support and socio-economic indicators, 1999

	Support per ha UAA	Support per AWU	GDP per head	Unemploy-ment rate	Population change 1989–99
	Based on FADN farm receipts				
Support per ha UAA	1	0.740(**)	0.143(**)	−0.244(**)	0.048
N	1063	1062	1063	957	892
Support per AWU	0.740(**)	1	0.004	−0.181(**)	0.034

Table 4.4 (cont)

	Support per ha UAA	Support per AWU	GDP per head	Unemploy-ment rate	Population change 1989–99
N	1062	1062	1062	956	892
	Based on appointed RD budget data				
Support per ha UAA	1	0.366(**)	−0.040	−0.095(**)	−0.024
N	1063	1062	1063	957	892
Support per AWU	0.366(**)	1	−0.106(**)	−0.061	−0.013
N	1062	1066	1066	960	892

** Correlation statistically significant at the 0.01 level (2-tailed).

The discrepancy between these measures could either be because LFA and agri-environmental payments are distributed more to richer areas while the remaining Pillar 2 measures are used more in poorer areas (see below), or because of a systematic difference between budgeted expenditure and actual payments received by farmers, with budgets underspent in poorer areas.

Various reasons have been proposed to explain why Pillar 2 instruments are not more supportive of the cohesion objectives of the EU. These include:

- differing national priorities,
- the uneven allocation of RDR funds, and
- difficulties co-financing RDR expenditure in poorer countries (Dwyer et al., 2002)

Agra CEAS (2003) also argue that the requirement of Member State co-financing of Pillar 2 measures has influenced uptake of the different measures. It follows that, even allowing for the different natural production conditions and the high variation in agricultural structures, the impact of Pillar 2 measures will be differentiated across EU space.

While the results above are based on an EU-wide analysis, a similar pattern holds in some Member States. For example, INEA (2002a) has mapped the regional distribution of product support in Italy and found a wide spatial variation, with the overall effect working clearly against cohesion objectives. Support per farm unit is highest in the northern areas, notably the favoured area of the Po valley, and in some central regions, and is least in the poorer south.

The changing nature and demands placed on rural areas means that the relationship between CAP support and prosperity of regions is unlikely to stay constant. For example, Lafferty et al. (1999) showed that in the 1970s the southeast of Ireland was regarded as one of the more prosperous regions, in part due to its strong agricultural sector. At this stage it also received high levels of CAP support. By the 1990s, while still in receipt of above-average CAP support, the region was no longer one of the most prosperous since its total economy has not adapted as well as other regions in Ireland. This highlights the limits of the analysis presented so far and the need to consider not only the incidence but the impact of CAP support (see Chapters 6 and 7).

Map 4.6 and Map 4.7 show the distribution of total Pillar 2 support on a per AWU basis.

Pillar 2 of the CAP comprises a number of quite distinct structural and rural development measures, as described in Chapter 2. For the EU as a whole, LFA and agri-environmental measures dominate Pillar 2 (Peters, 2002). However, the relative importance of different Pillar 2 measures varies widely between Member States, reflecting amongst other things different national priorities and different national budget constraints. The territorial distribution of support through each of these categories is therefore considered separately in the following subsections.

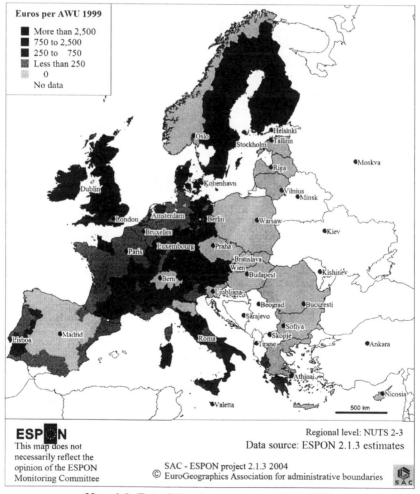

Map 4.6: Total Pillar 2 support per AWU, 1999

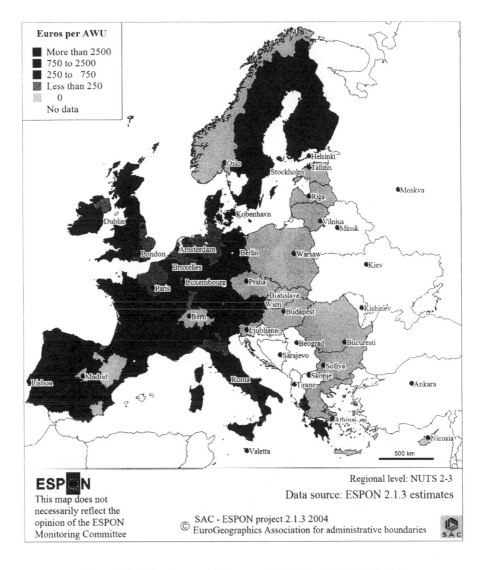

Map 4.7: Pillar 2 expenditure per AWU (from RDR budgets)

4.6. LFA payments

Empirical analysis of the distribution of LFA support was based on data from the European Court of Auditors (ECA, 2003) concerning rural development: support for less-favoured areas, together with the Commission replies and apportioned to NUTS3 regions according to the method described above.

Given the objectives of the LFA scheme, it was expected that the scheme would operate in a manner consistent with the economic cohesion objectives of the EU. In other words, LFA payments would tend to be higher in regions with lower per capita GDP and higher unemployment rates. Correlation analysis, however, found no statistically significant relationships between levels of LFA support and indicators of economic cohesion, although the signs of the coefficients were as expected (Table 4.5). LFA support therefore is only weakly related to the indicators of social and economic cohesion. Both Peters (2002) and Agra CEAS (2003) discuss how the profile of different accompanying measures varies considerably between Member States, with certain richer northern States (including Finland, France and Luxembourg) prioritising the LFA scheme over agri-environmental, farm investment or early retirement schemes. This may explain why a stronger relationship with the cohesion indicators was not detected.

Table 4.5: Pearson correlation coefficients between level of LFA payments and socio-economic indicators

	Support per ha UAA	Support per AWU	GDP per head	Unemploy-ment rate	Population change 1989–99
Support per ha UAA	1	0.955(**)	−0.011	0.043	−0.045
N	1063	1062	1063	957	892
Support per AWU	0.955(**)	1	−0.055	0.057	−0.037
N	1062	1068	1068	962	894

** Correlation is significant at the 0.01 level (2-tailed).

4.6.1   Agri-environmental payments

As described in Chapter 2, the CAP reforms of the early 1990s included the introduction of a number of accompanying measures, of which agri-environment schemes were the most notable. Quantitative analysis of the territorial incidence of support through agri-environmental schemes was based on data from the FADN database showing the value of environment-related payments received by farmers. As previously, correlation analysis was carried out to assess whether the level of agri-environmental support is distributed in a manner consistent with EU cohesion objectives (see Table 4.6).

Table 4.6: Pearson correlation coefficients between level of agri-environmental subsidies and socio-economic indicators

	Support per ha UAA	Support per AWU	GDP per head	Unemploy-ment rate	Population change 1989–99
Support per ha UAA	1	0.801(**)	0.146(**)	−0.240 (**)	0.046
N	1067	1066	1067	961	894
Support per AWU	0.801(**)	1	0.017	−0.158(**)	0.014
N	1066	1069	1069	963	894

** Correlation is significant at the 0.01 level (2-tailed).

The results show that higher levels of agri-environmental payments in 1999 accrued to richer areas of the EU. In other words, the distribution of agri-environmental payments was not consistent with economic cohesion objectives. The findings reflect the fact that richer EU Member States tend to prioritise agri-environmental objectives more than poorer regions. Sweden and Austria, for example, allocate over 50% of their total RD funding to the agri-environment measures (Peters, 2002).

The only national-level study identified which considered the consistency between agri-environmental schemes and cohesion objectives related to Ireland (Lafferty *et al.,* 1999). In this case the authors argue that agri-environmental schemes are contributing to the achievement of the economic and social cohesion goals and helping to constrain tendencies towards abandonment of farmland.

Map 4.8 and Map 4.9 show the distribution of these two elements of Pillar 2 of the CAP – LFA support and agri-environmental subsidies. Both are expressed on a per AWU basis. It can be seen that each tends to favour the northern European and Alpine regions of Europe, with less use in southern Europe.

4.6.2   Rural development measures

The extent to which any benefits of rural development schemes are territorially specific depends on whether programmes are themselves spatially oriented (such as Objective 5b and LEADER). However, it was expected that that incidence (and thus potential impact) of structural expenditures would be territorially focused even when the programmes are not, as the take-up rates are usually variable across farming types or scales of farming, which in turn are regionally specific. Unfortunately statistical analysis of the territorial incidence of rural development measures was constrained by a lack of information on policy expenditures. Instead, Chapter 5 explores in detail the take-up and impact of the LEADER scheme and other Article 33 measures.

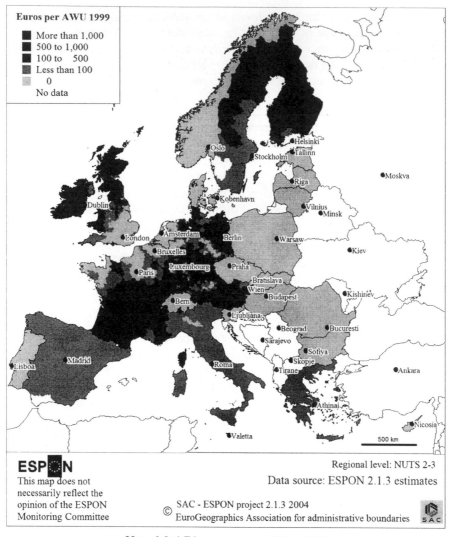

Map 4.8: LFA support per AWU, 1999

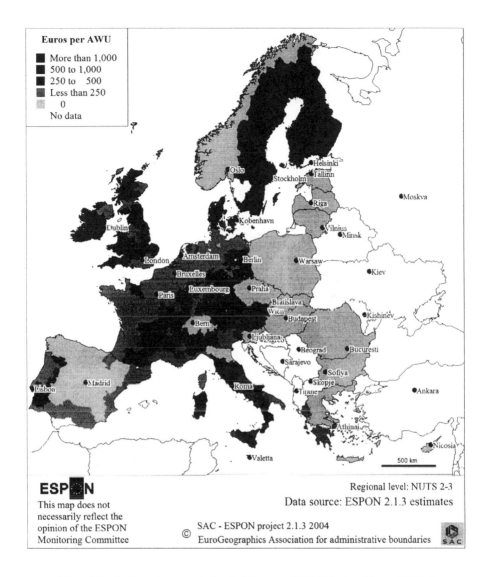

Map 4.9: Agri-environmental subsidies per AWU (derived from FADN)

4.7. The influence of farm type

Considering the relationship between farm size and CAP support, it is widely recognised that CAP Pillar 1 support accrues disproportionately to intensive large-scale farmers. This is because to date it has been coupled (either directly or indirectly) to the level of output produced by a farmer. However, whether this leads to support *per hectare* being higher for larger farms is harder to predict.

Table 4.7 reports the findings from a cross-tabulation of Pillar 1 support per ha UAA against the average farm size in each region measured in European Size Units (ESUs). Information on the latter was taken from the FADN database apportioned to NUTS3.

Both CAP support and farm size were banded into five groups on the basis of quintiles. Thus, the regions with support levels smaller than or equal to the first quintile formed the first group, the second group comprised those NUTS3 with values above the first quintile and smaller than or equal to the second quintile, etc. As previously, analysis focuses on 1999.

Table 4.7: Cross-tabulation of per hectare Pillar 1 support and farm economic size, 1999

Pillar 1 support per ha classification (1 = lowest)	Farm size classification (1 = smallest)					
	1	2	3	4	5	Total
1	89	45	47	9	20	210
	42.4%	21.4%	22.4%	4.3%	9.5%	100.0%
2	26	32	56	23	73	210
	12.4%	15.2%	26.7%	11.0%	34.8%	100.0%
3	32	39	51	41	46	209
	15.3%	18.7%	24.4%	19.6%	22.0%	100.0%
4	43	68	26	49	24	210
	20.5%	32.4%	12.4%	23.3%	11.4%	100.0%
5	20	26	30	87	46	209
	9.6%	12.4%	14.4%	41.6%	22.0%	100.0%
Total	210	210	210	209	209	1048

Chi-square tests of joint association confirm a relationship between the distribution of Pillar 1 support per ha and average farm size: regions with larger farms tend to receive higher levels of CAP support per ha UAA. In particular, Table 4.7 shows that, in 1999, 42% of regions receiving the lowest level of support fell into the smallest farm size category while 64% of those regions receiving the highest level of support fell into the two largest average economic size categories.

Since large farms are mainly concentrated in continental Europe while small farms are more prominent in southern regions of the EU, the results reported in Table 4.7 suggest an uneven territorial distribution of Pillar 1 support across Europe. Differences in the distribution of support have also been detected within country-level studies. For example, in Germany there are large differences between different Laender regarding the average farm size. In 1997, farm size differs between averages of 24.7 ha in the so-called Alte Laender and 201.7 ha in the Neue Laender and between 17.9 ha in the Land "Baden-Wuerttemberg" (in the southwest) and 272.2 ha in "Mecklenburg Vorpommern" (in the northeast) (Statistisches Bundesamt, 1999). The role of CAP in regions with larger farm sizes, such as "Mecklenburg Vorpommern", is much higher than in regions with smaller sizes of farms, such as "Baden-Wuerttemberg".

The same cross-tabulation analysis was repeated for Pillar 2 support. As shown in Table 4.8, the distribution of Pillar 2 support was found to be much less strongly related to farm size. In this case, a very large proportion (77%) of those in the highest support group fell into the smallest two farm size categories.

Table 4.8: Cross-tabulation of Pillar 2 CAP support measures in relation to farm economic size, 1999

Level of Pillar 2 support	Farm size classification (groups)					
	1 (Smallest)	2	3	4	5 (Largest)	Total
1 (lowest)	44	8	35	81	44	212
	20.8%	3.8%	16.5%	38.2%	20.8%	100.0%
2	10	12	43	80	67	212
	4.7%	5.7%	20.3%	37.7%	31.6%	100.0%
3	35	25	69	34	49	212
	16.5%	11.8%	32.5%	16.0%	23.1%	100.0%
4	50	76	43	12	31	212
	23.6%	35.8%	20.3%	5.7%	14.6%	100.0%
5 (highest)	73	91	22	5	22	213
	34.3%	42.7%	10.3%	2.3%	10.3%	100.0%
Total	212	212	212	212	213	1061

Note: Estimates of Pillar 2 support in this case were based on the apportioned funds for RD measures. The alternative, FADN-based estimates of Pillar 2 support revealed a similar relationship.

The level of support not only varies by farm size but also between commodities and, in general, does not differentiate between production conditions or production methods. Taking this into account, Table 4.9 reports the results from an analysis of the relative importance of different factors on the distribution of support across Europe. The key explanatory variables investigated were farm size, land cover (as a proxy for the type of farm and productive capability of the land) and intensity of production as measured by Farm Net Value Added per ha.

Table 4.9: Agricultural factors influencing the level of CAP support

	Pillar 1 support per 100 ha		Pillar 2 support (FADN) per 100 ha		Pillar 2 support (RDP) per 100 ha	
	β	t	β	t	β	t
(Constant)	46.638	5.005	10.969	10.360	10.317	7.347
Average farm size	0.682	11.746	−0.079	−12.016	−0.055	−6.273
% Irrigated	2.425	4.804	−0.196	−3.417	−0.228	−2.998
% Rice	0.399	0.374	−0.052	−0.430	−0.017	−0.106
% Viniculture	−1.117	−4.850	−0.029	−1.092	0.081	2.331
% Fruit	−2.151	−4.552	−0.032	−0.590	0.137	1.922
% Olives	−0.331	−1.071	0.031	0.869	0.085	1.816
% Permanent pasture	0.227	3.667	0.005	0.671	−0.017	−1.827
% Assoc. crops	0.908	1.531	−0.011	−0.170	0.240	2.682
% Small parcels	0.906	8.882	−0.003	−0.219	0.046	2.966
% Natural vegetation	0.464	2.575	−0.093	−4.531	0.036	1.340
Intensity of production	0.006	7.909	0.000	1.190	0.000	1.646
GDP per head	0.000	−0.353	0.000	−0.526	0.000	0.639
Unemployment rate	−1.911	−4.274	−0.293	−5.773	−0.321	−4.762
Population change	1.680	0.273	−0.155	−5.007	−0.145	−3.532
R^2 etc.						

The results show that, controlling for other factors, average farm size is a significant factor in explaining the level of CAP support received by NUTS3 regions. In the case of Pillar 1 support, regions with larger farms get higher levels of support. In contrast, higher levels of Pillar 2 support tend to go to regions with smaller average farm sizes.

Turning to the land cover variables, as anticipated, these are shown to play a more significant role in explaining the distribution of Pillar 1 support than they do for Pillar 2 support. For example, six of the land cover types included in the analysis were significant factors in explaining Pillar 1 support per hectare UAA, as compared to only two which were significant at the 5% level in explaining Pillar 2 support per hectare UAA. In the case of Pillar 1 support the percentage of land accounted for by irrigated land and permanent pasture had positive and significant coefficients while negative coefficients were estimated for the percentage of land covered by vine or fruit. These findings reflect the varying level of MPS given for different agricultural products. The intensity of production is, *ceteris paribus*, positively related to Pillar 1 support levels but appears to be less important in determining the Pillar 2 received by a region.[15]

Importantly, the level of per capita GDP is shown to become an insignificant influence on the level of both Pillar 1 and Pillar 2 support once other factors are accounted for. In other words, the relationships shown in Table 4.1 and Table 4.2 above can, at least in part, be explained by the way in which a combination of farm-related variables vary across space. However, the coefficients associated with unemployment rates and, in the case of Pillar 2, population change continue to suggest that the distribution of support is not entirely consistent with higher-level EU cohesion objectives.

The review of other country-level studies supports the hypothesis that the impact of the CAP is highly differentiated by farm type. For example, in Austria, average support levels are highest for farms specialising in field crops that are concentrated in the lowlands, particularly in northeast of Austria. These farms have support levels at least 50% higher than all other types of farming (BMLFUW, 2000). Similarly, Lafferty *et al.* (1999) show that in Ireland the greatest impacts of the CAP occur in the southeast and southwest where, on average, farms are largest and there is a higher level of specialisation on dairying and cereals.

4.8. The influence of accessibility and regional type

4.8.1 Accessibility

Consistent with other ESPON projects, the distribution of CAP support in relation to three territorial scales of accessibility were considered – the macro or EU-wide level, the meso level and micro or local level (see Table 4.10). In each case, the lower the value of the indicator, the greater the accessibility of the region.

Table 4.10: Indicators of accessibility

Scale	Source	Description
Macro	ESPON Database Vn. 2.3 (2.1.1_Timetomarket_Access-ibility_by_rail_road_N3)	Accessibility time to market by rail and road, half-life (1000 minutes), weighted by GDP (1997)
Meso	ESPON Database Vn. 2.3 (2.1.1_Timetomarket_Access-ibility_by_rail_road_N3)	Accessibility time to market by rail and road, half-life (25 minutes) weighted by GDP (1997)
Micro	ESPON Project 1.2.1. Mcrit. (ICON_access_transport_ter-minals_2001)	Accessibility by road to transport terminals offering a minimum service

---

[15] Subsequent analysis found no statistical evidence of a relationship between changing intensity of production and support levels.

An alternative peripherality indicator, developed for the European Commission by Schurmann and Talaat (2000), was also used in the analysis to test the sensitivity of the findings. While focused like the macro indicator at the EU level, this indicator was based purely on road accessibility to EU-15 centres as opposed to market potential. In this case, the lower the value of the indicator, the more peripheral the region, and *vice versa*. One would therefore expect the two EU-scale indicators to give rise to similar results but with opposite signs.[16]

Correlation coefficients between the level of Pillar 1 support received and the accessibility of each NUTS3 region at each spatial scale are presented in Table 4.11 below.

Table 4.11: Pearson correlation coefficients between level of total Pillar 1 support and accessibility indicators

	Accessibility indicators[1]			
	Micro	Meso	Macro	Peripherality index[2]
Support per ha UAA	−0.374 (**)	−0.293(**)	−0.251(**)	0.382(**)
N	1044	1047	1047	1047
Support per AWU	−0.119(**)	−0.035	−0.295(**)	0.232(**)
N	1046	1049	1049	1049

** Correlation is significant at the 0.01 level (2-tailed).
[1] The lower the value of the indicator, the greater the accessibility of the region.
[2] The lower the value of the indicator, the greater the peripherality of the region.

The results suggest that the level of Pillar 1 support, both per hectare and per AWU, tends to be higher in more accessible regions, and lower in more peripheral regions, at all spatial scales. All but one of the correlation coefficients is significant. From a spatial policy perspective, these findings confirm that although Pillar 1 measures are aspatial, they have very discernible spatial impacts.

Table 4.12 and Table 4.13 present the findings in relation to the spatial distribution of Pillar 2 support, the first based on FADN data, the second apportioned Rural Development Plan budget data.

Table 4.12: Pearson correlation coefficients between the level of total Pillar 2 support as estimated from FADN data and accessibility indicators

	Accessibility indicators			
	Micro	Meso	Macro	Peripherality index
Support per ha UAA	0.244(**)	0.103(**)	−0.017	−0.152(**)
N	1056	1059	1059	1059

[16] Different results arise because the indicators focus on different aspects of accessibility. Personal communication with members of EPSON project 1.1.2 revealed that other EU-level indicators of accessibility are being developed as part of the EPSON programme.

Table 4.12 (cont)

	Accessibility indicators			
	Micro	Meso	Macro	Peripherality index
Support per AWU	0.359(**)	0.259(**)	0.050	−0.189(**)
N	1055	1058	1058	1058

** Correlation is significant at the 0.01 level (2-tailed).

Table 4.13: Pearson correlation coefficients between the level of total Pillar 2 support as estimated from RDP budget data and accessibility indicators

	Accessibility indicators			
	Micro	Meso	Macro	Peripherality index
Support per ha UAA	0.191(**)	0.188(**)	0.189(**)	−0.222(**)
N	1056	1059	1059	1059
Support per AWU	0.468(**)	0.389(**)	0.142(**)	−0.298(**)
N	1059	1062	1062	1062

** Correlation is significant at the 0.01 level (2-tailed).

The results in relation to Pillar 2 support are very different to those of Pillar 1. In particular, they suggest that the least accessible regions received, on average, higher levels of Pillar 2 support. Similarly, higher levels of Pillar 2 support were found in more peripheral regions, again at all spatial scales, local, meso and macro.

Various hypotheses were considered by the project team in relation to the effects of the CAP on spatial patterns of development. For example, it was argued that changes in the levels of farm household pluriactivity may be more strongly associated with variables reflecting the strength of the local economy than the level of CAP support. Regression analysis backed this up with findings that the relationship between the level of CAP support received by a region and the extent of part-time farming was not statistically significant.

4.8.2 Regional types

The distribution of CAP support across different types of NUTS3 regions in Europe was analysed using both the OECD and the urban-rural regional typologies. Table 4.14 considers the first of these, and indicates how the share of total support received by each OECD-classified region compares to their share of AWUs and UAA.

Table 4.14: The incidence of CAP support by OECD region

OECD region type[1]	% of total			UAA (%)	AWU (%)
	Pillar 1	Pillar 2 (FADN)	Pillar 2 (RD)		
Rural – leading	18	22	20	20	15
Rural – lagging	27	29	26	31	22
Intermediate – leading	14	13	13	12	12
Intermediate – lagging	18	18	21	20	23
Urban – leading	12	7	7	8	12
Urban – lagging	10	7	5	5	8
Missing	1	4	9	3	8
TOTAL	100	100	100	100	100

[1]Based on data supplied by OECD

The results show that, as expected, predominantly rural regions receive the lion's share of total CAP support with predominantly rural regions receiving 45% of total Pillar 1 support and either 50 or 46% of Pillar 2 support, depending on whether the FADN or apportioned RDR fund data is used as the basis for the analysis. Rural regions receive a substantially higher share of both total Pillar 1 and Pillar 2 support than their share of the agricultural work force would suggest.

The findings when the urban-rural typology is used are more surprising. Here, regions labelled as rural but with either high or medium human intervention account for only a small percentage of total EU agricultural area (4 and 6% respectively) and total agricultural workforce (7 and 4% respectively). Their share of Pillar 1 and 2 support is consistent with this, i.e. very low. In contrast the categories labelled "urban, high human intervention" and "urban, low human intervention" account for the most significant shares of agricultural area and labour force and also receive the largest shares of both types of support. The fact that 44% of total Pillar 1 support accrues to the urban, high human intervention category of regions is particularly noticeable (Table 4.15).

Table 4.15: The incidence of CAP support by degree of urban integration

Regional type[1]	% of total			Total UAA(%)	Total AWU (%)
	Pillar 1	Pillar 2 (FADN)	Pillar 2 (RD)		
Rural, high human intervention	4	5	6	4	7
Rural, medium human intervention	8	4	3	6	4
Rural, low human intervention	12	25	30	21	22
Urban, high human intervention	44	34	23	29	35

Table 4.15 (cont)

Regional type[1]	% of total			Total UAA(%)	Total AWU (%)
	Pillar 1	Pillar 2 (FADN)	Pillar 2 (RD)		
Urban, medium human intervention	1	4	5	2	6
Urban, low human intervention	30	28	32	37	25
Missing	1	0	1	0	1
TOTAL	100	100	100	100	100

[1]Based on regional typology derived by ESPON project 1.1.2

Specifically in relation to the role of CAP and the diversification of rural areas, the project team hypothesised that changes in the levels of farm household pluriactivity are more strongly associated with variables reflecting the strength of the local economy than the level of CAP support.

In general, the literature review indicated that rates and patterns of farm household pluriactivity vary widely both between and within Member States and the CAP ranks fairly low as a factor driving change (Edmond and Crabtree, 1994; Cawley *et al.*, 1995; McDonagh *et al.*, 1999). Regression analysis backed this up with findings that the relationship between the level of CAP support received by a region and extent of part-time farming was not statistically significant. Interestingly, the strength of the regional economy was found to be negatively related to rates of part-time farming, in other words, part-time farming was more prevalent in poorer regions of the EU than richer regions.

Chapter 5

5. Adjustments and Impacts of the CAP/RDP

5.1. Integrating experiences from regional contexts

With rising surpluses in agricultural production of industrialised countries, agricultural ministries have broadened the orientation of their policies beyond the farm sector to include improvement of economic opportunities in rural areas, the sustainability of the natural environment and the provision of countryside amenity. This more or less common trend in European countries has been supported by the assessment of farm household behaviour and analysis of an increasing involvement in diversification and off-farm activities.

At first, attention focused on the (long-established) phenomenon of combining *agricultural and non-agricultural activities of farm households*. Taking the farm unit as starting point of observation, this combination was first referred to as *part-time farming*, and then shifted to an understanding of *multiple-job farm households*. The rise of the term *pluriactivity* relates to the diversity of farm households and adjustment strategies observed. This diversity has gradually gained general acceptance, e.g. among agricultural policy-makers and (partly) interest groups. With more detailed studies on the scope, type, extent and tendencies of these combinations of activities and situations of farm households at the end of the 1980s and in the 1990s (e.g. Bryden *et al.*, 1993; Commins and Keane, 1995; Dax *et al.*, 1995; Bowler *et al.*, 1995), it was understood that it was not simply the existence of different jobs which was essential for farm households but that of complementary activities of varying extent and at different places.

A clear understanding of structural adjustment in agriculture is fundamental for effective policy design in rural areas. The structural characteristics of the sector and changes in these structures have an important bearing on the viability of the rural economy as they affect the magnitude and distribution of income and economic activity and have spatial consequences. The relative importance of the agricultural sector can vary dramatically by location between and within countries according to the different degrees of economic diversification in rural areas (OECD, 1998b).

There is a considerable variation among farm households in the share of their labour allocated to farming and non-farming activities, and, to an even greater extent, in the dependence of farm households on incomes from agriculture. Despite a general growth in off-farm work over past decades, there is a wide range of contextual factors, in particular region-specific social and cultural elements, shaping the actual adjustment patterns. Engagement in off-farm work can have an important role with regard to agricultural policy reform, cushioning farm households from income pressures. Many farm households, particularly in less-favoured areas, are dependent on single or very limited farm production sources for their incomes. By enabling farm households to diversify their income sources, pluriactivity can contribute to diversification and lower exposure to farm-sector events. Of course, the regional economy and the availability of non-farm employment opportunities largely impacts on these options.

Many factors such as region, the structural characteristics of the farm and the household, and the economic environment, in particular the opportunities for off-farm work, affect the total income of farm households. The inclusion of off-farm income narrows the dispersion of income by farm size and farm type, and the total income of farm households is therefore more equally distributed than that of farm income. The same holds true for the analysis of income differences by region. Basically, these differences arise from regional variations in the economic size of farms, type of farming and rate of support for each commodity, and depend on how widely regions are defined. For example, income differences across regions in Denmark are less than across farm types or size classes. In Switzerland, the average farm income in lowland areas is 11% higher than the average of all farms while that of mountain farms is 21% lower (OECD, 2003b). These findings probably also apply to the LFA regions of the European Union. When non-agricultural incomes are taken into account, regional differences in income across farms of different size and type are reduced.

When assessing the territorial impact of agricultural and rural development policies through analysis of household adjustment strategies, it is necessary to keep a number of considerations in mind. First, policies differ in the extent to which they have an explicit territorial dimension; measures for less-favoured areas, for example, will obviously differ from market price supports. Second, there are territorial differences in resource endowments. Third, farm structures (farm sizes and distribution), infrastructures and capacity to adopt innovations are often poorer or more restrictive in those same regions where land resources are less productive. Fourth, there can be regional variations in historical, institutional or cultural factors and in the manner in which these underline farm traditions and practices. Attitudes to farmer retirement and property transfer show clear regional differences (Lafferty *et al.,* 1999). Fifth, layered over these structural features is a complex of dynamic economic and technological forces which seem to have universal application in the way in which they "drive" the longer-term pathway of agricultural restructuring and adjustment. This longer-term trajectory is characterised by inelastic demand for agricultural products, constant downward pressure on farm-gate prices, labour outflows from farming, enlargement of farm business scale, polarisation of farm incomes between commercial producers and marginalised categories of producer, reliance on "non-market" subsidies, and dependence of farm households on non-farm income sources (Commins and Keane, 1995). This last point bears out the increasing importance of economic development in rural areas, and the significance of non-farm enterprise and employment in facilitating farm household adaptations to the declining economic viability of farm businesses. It also signals the importance of enabling rural communities to access urban networks and inter-urban communication routes (McHugh and Walsh, 2000).

The overall assessment of policy impact is rendered particularly difficult through the numerous separate policy instruments as parts of the CAP, and its inter-connectedness with other policy, economic and socio-economic aspects. Following on the MacSharry reform of 1992, a continuing discussion on CAP reform and several rounds of reforms have led to quite substantial changes in the policy framework. Since the mid-1990s, the need for a shift from the CAP to an integrated rural policy has been clearly formulated, and a model for such an alternative policy framework was put forward by the group of experts charged by the European Commission to outline the principles that might guide the transition of the CAP towards the integration of environmental and rural development objectives (Buckwell *et al.*, 1997). The stepwise transition presented in that report is still seen frequently as the reference for current policy reform.

Agenda 2000 and the reforms of July 2003 have made contributions towards integration and a stronger focus on rural development policy. Yet, implementation of the RDP schemes has been rather weak, and the Rural Development Programmes elaborated after Agenda 2000 have brought only a very limited increase of rural development funds in most EU countries. Initially high expectations among some stakeholders, e.g. environmental organisations and rural development groups, for radical reform of the CAP into a policy for sustainable rural development, in which the RDP would play a key role, have been disappointed by the slow pace of reform and by strong political resistance to any move that could be seen as taking money away from farmers (Lowe and Brouwer, 2000). In reality, it has turned out that the Rural Development Programmes are focused on support for activities "close to agriculture". There are, however, striking differences in the patterns of RDP (and also SAPARD) expenditure. These broadly reflect historical allocations to similar measures in the past and have not been fundamentally altered by the new Programmes (Dwyer *et al.*, 2002).

5.2. Case study methods

The rest of this chapter reviews and synthesises a number of case studies of the regional impact of CAP/RDP, using information from available literature. Many relevant studies at regional, national and European levels have been carried out, and official evaluation studies cover increasingly all types of CAP instruments. However, most of these studies do not include region-specific information and explicit spatial impact analysis. Other reference sources used are the reports of research projects (primarily international), some carried out within the EU's Research Framework Programmes. In particular, use has been made of those projects which address region-specific application of CAP measures or which focus on institutional aspects of rural development processes.

The case studies were undertaken to provide deepened insight into the core issues, i.e. detailed empirical information on the territorial incidence of the CAP/RDP, and more evidence on the impact of CAP/RDP measures on the economic, social and environmental situation in regions of different types. The focus is on more differentiated territorial information about the application of CAP/RDP instruments, in particular the territorial (mainly regional) effect of specific schemes, and/or the combined effect of policy programmes.

Case studies were selected on the basis of a number of criteria, i.e.:

- availability of information, especially pre-existing and relevant studies and evaluations
- quantitative assessments at national and EU level
- application, i.e. the scope of measure(s) in national contexts where measures are applied horizontally, or regional application, geographical definition, "quality" of application, etc.
- cluster results (see Shucksmith *et al.*, 2004).

The policy instruments selected for case study are four major sub-programmes/schemes of CAP Pillar 2, as implemented in many EU-15 countries and in Norway; these are listed in Table 5.1. Where possible, analysis of "good practice" (which is taken to include "good structures"[17]) by RDP organisations was undertaken, e.g. the institutional context and behaviour for Pillar 2 measures, such as take-up rates, eligibility, consultation, advisors, co-funding, support structures, etc.

In addition, the SAPARD programme designed as a pre-accession instrument for countries in Central Europe is studied, and two countries – Ireland and (pre-accession) Poland – have been singled out for case-study treatment at a broader level. The Irish study takes a number of contemporaneous economic developments and support schemes into account, while the Polish one examines the practical details of implementing the SAPARD programme in that country.

Table 5.1: Case study instruments

CAP (Pillar 2) instrument	Case studies
Agri-environmental programme (AEP)	Spain (Steppeland, Castilla y Léon; Castilla La Mancha; Guadalquivir river) Germany Austria (Bludenz-Bregenzerwald) Ireland Hungary Norway

[17] That is, the shape, size etc. of the organisations and agencies which promote rural development.

Table 5.1 (cont)

CAP (Pillar 2) instrument	Case studies
Farmers' early retirement scheme (ERS)	Greece (Lesvos)
	France
	Ireland
	Spain (Castille and Leon)
	Finland
	Norway
Less-favoured areas scheme (LFA)	Austria (Bludenz-Bregenzerwald)
	UK (Scotland)
	Greece
	Sweden
	Slovenia
	Hungary
LEADER programme	Austria (Bludenz-Bregenzerwald)
	Germany (with Regionen Aktiv)
	Spain (Adema, in Soria)

5.3. Agri-Environmental Programmes (AEPs)

5.3.1   Introduction

The structural transformation of agricultural production throughout Europe in the second half of the twentieth century has, on balance, contributed to a number of environmental problems. Technological and economic changes have resulted in increased levels of intensification, specialisation and concentration, which in many areas resulted in negative externalities that include ecological effects such as a reduction in biodiversity and loss of habitat and landscape features, as well as growing problems of soil degradation, water depletion and contamination, and also air pollution (CEC, 1998; Baldock *et al.*, 2002; OECD, 2003b). There is evidence of considerable dysfunctionality in terms of negative consequences generated for the rest of society by agricultural production (Matthews, 2002). The abandonment of farmland also creates pressure on the countryside and biodiversity (CEC, 1999).

The relative importance of different environmental issues depends on the effects of farming practices at different geographical scales – local, regional, national or international – which in turn reflect variations in climatic and ecological factors. In addition, the level of priority attached by society to agriculture-induced environmental issues can also vary between and within regions, depending on such factors as population density, income levels and the value attached to cultural heritage (OECD, 2003b).

Policy instruments have in many cases exacerbated the environmental problems associated with the modernisation of agriculture. The CAP has been predominantly focused on assisting farmers to intensify their production systems, including expansion of production onto environmentally sensitive or marginal areas. While it is difficult to establish precisely the contribution of agriculture policy to environmental damage, there is little doubt that the CAP has exacerbated the negative environmental externalities associated with modernisation of farming in many regions (Baldock *et al.*, 2002). For example, in Germany, it is estimated that:

- 29% of fern plants and flowering plants (Farn- und Bluetenpflanzen)
- 36% of bird species (Vogelarten)
- 47% of aboriginal mammal species (einheimischen Saeugetierarten)
- 58% of amphibian species (Lurcharten)

have been lost as a result of the intensification of agricultural production methods (Loesch and Meimberg, 1986; Voegel, 1993). Similarly, in Ireland, the range of bird species has been reduced in areas of intensive farming: the most commonly cited example is the retreat of the corncrake following the switch from hay to silage as winter fodder for livestock.

The impacts on landscape quality that have been associated with intensification include:

- removal of field boundaries (which also includes loss of habitats for flora and fauna);
- destruction of archaeological monuments; and
- detrimental impacts on the environment such as pollution of river water, eutrophication of lakes, and significant contributions to methane gas emissions (Stapleton *et al.*, 2000).

A study of the financial costs of UK agriculture as a whole in 1996 found that the total cost of all environmental externalities (including those associated with human health) is equivalent to 13% of the total average gross income of the sector in the 1990s (Pretty *et al.*, 2000). As with other studies, very different environmental costs were associated with different types of farming. Given the distinct geographical pattern of farm types across Europe, this finding supports the hypothesis that negative environmental effects of agriculture are territorially specific. However, the extent to which these effects can be specifically attributed to the CAP is less clear (Baldock *et al.*, 2002).

5.3.2 AEP objectives

Since the ratification of the Maastricht Treaty, the EU has a legal obligation to take account of environmental protection requirements when drawing up and implementing Community policies, an obligation which was reinforced by the Amsterdam Treaty of May 1999. The ESDP has as one of its core objectives sustainable development, prudent management and protection of nature and cultural heritage (EC, 1999). These problems may be seen (CEC, 1998) as examples of market failure, in that private incentives are inadequate to prevent over-supply of negative "externalities" such as pollution, or the degradation of "public goods" such as wildlife and landscape beauty.

With Council Regulation no. (EEC) 2078/92, the 1992 reform of the CAP introduced support measures for agri-environment measures at European level to encourage more environment-friendly production methods. By the end of the 1990s, coverage under agri-environmental payment contracts reached almost 20% of farmland in the EU (OECD, 2003b), and agri-environmental programmes as part of the suite of measures that make up the Rural Development Regulation introduced as part of the Agenda 2000 reforms now account for approximately 30% of total rural development funds in the EU-15. Agri-environment measures also form a significant element in the Rural Development Programmes of many of the new Member States, with up to 44% of the total Programme budget in Hungary and amounts in excess of 30% in some other states (see WWF, 2004). However, on average, the share of total Guarantee Section rural development funds is only 18% (preliminary indicative budget).

Member States are required to apply agri-environment programmes throughout their territories, according to environmental needs and potential. The objectives of such programmes fall into two broad categories:

- To reduce the negative pressures of farming on the environment, in particular on water quality, soil and biodiversity;
- To promote farm practices necessary for the maintenance of biodiversity and landscape, including avoidance of degradation and fire risk from under-use.

It has been argued (Schramek *et al.,* 1999) that there is a lack of clearly specified environmental objectives in the scheme as established by the EU, and the majority of measures applied in Member States are mainly focused on agricultural practices. It has also been argued (Harte and O'Connell, 2002) that agri-environment payments function as income supports conditional upon delivering environmental benefits (the cross-compliance model) rather than as payments for environmental outputs.

Environmental outcomes related to agricultural practices are not limited to the agri-environment Regulation but are also addressed through the Birds and Habitats Directive, the Water Framework and Nitrates Directives, and associated regulations. In addition, some specific measures supported under the Structural Funds (ERDF) counteract the effects of intensive farming, for example the National Scheme for Control of Farmyard Pollution in Ireland.

The main elements characterising agri-environment agreements include the following:

- farmers deliver an environmental service;
- agreements are voluntary for the farmers;
- measures apply only on farmland;
- payments cover income foregone, costs incurred and a necessary incentive.

In addition to the land management measures, the 1992 Regulation provided for training and demonstration projects to promote the use of environmentally beneficial techniques and good farming practice. Regional or national authorities manage the programmes under a decentralised system of management, subject to approval by the Commission for each programme. Administration is normally undertaken by the agriculture authorities, with the environmental authorities often responsible for programme development, implementation, monitoring and evaluation. In a few cases, the environmental authorities manage the programmes.

A flexible administrative framework, encouraged under the Council legislation, has led to a variety of programme structures in the Member States, including the countries that have most recently joined the EU. The majority of Member States have adopted zonal programmes, established at different administrative levels (national, sub-national, and regional) and normally including general measures which concern all qualifying farmers in the administrative territory and more specific schemes for designated zones. In addition, most measures are only applicable to certain types of crop or land use. Most Member States have adopted horizontal measures through all the national territory, particularly for organic farming and training programmes. In Spain, for example, there are horizontal measures applicable to the whole country covering extensification, breed and strain preservation, organic farming and agri-environment training.

Different priorities and concerns of the first wave of agri-environment programmes have been identified at national and sub-national level (Schramek *et al.*, 1999). These comprise:

- a focus on nature and landscape protection and on mechanisms for changing agricultural land management (in the UK in particular; there are also strong naturalist traditions in Germany, Sweden and Austria);
- the economic support of marginal agricultural activities threatened by the abandonment of farming and compensation for natural handicaps (e.g. southern France, parts of Spain, Portugal and much of Greece);
- farm-based pollution (e.g. in Germany and Denmark);
- agricultural modernisation and structural reform (in southern European countries in particular).

Member States have as a consequence used different criteria to target programmes and measures grouped primarily under agricultural and environmental objectives, and this has contributed to some variation in take-up rates. While some countries (Finland, Austria, Sweden, Germany) have above-average rates, others (Belgium, Denmark, Greece, Spain and the Netherlands) fall below. Participation is generally higher among farmers engaged in more extensive farming systems and frequently on smaller farms (but not exclusively, e.g. Spain; Paniagua Mazorra, 2001). Schramek *et al.* (1999) have identified financial incentives as the most important motivation.

Among the reasons put forward to explain relatively low take-up rates are the innovative nature of the measures, their complexity, the problems caused for certain administrations, political priorities, the balance in certain Member States between central and regional governments, budgetary difficulties in certain Member States (or regions) in providing the necessary part-financing (in Hungary, for example, the number of applications in 2002 was 5321 but only

2691 were selected because of limited funds), the cultural reticence of some farmers and the economic benefits of continuing to practise intensive agriculture. There is also evidence that commodity supports actively discourage take-up of agri-environment measures (Brouwer and Lowe, 1998).

Within the CAP, the agri-environment programmes are (or were) innovative in many respects, including the importance given to subsidiarity (Member States draw up their own programmes), the fact that the participation of farmers in the programmes is voluntary, and the multi-annual nature of the programmes.

Regional production conditions influence the territorial impact of agri-environmental schemes. Firstly, farmers are only eligible for some agri-environmental schemes if they satisfy certain habitat-specific conditions and the ability to meet these conditions varies spatially. Secondly, the opportunity costs associated with complying with the conditions of agri-environmental schemes will vary across space as well as between farm types. Compensation payments are generally not sufficient to encourage intensive farmers to participate.

A number of studies have pointed to evidence of environmental improvements generated by the programmes including reduction in soil erosion and pollution, limiting pressure from input use, conservation of habitats and maintaining cultural landscapes (see for example CEC, 1998; Baldock *et al.*, 2002). There is strong evidence from Ireland of improvements in farming practices leading to reduced environmental impact (Teagasc, 2003) but evidence of positive impacts on biodiversity is more limited and indeed the application of some measures is possibly detrimental to biodiversity (Feehan *et al.*, 2002). However, the effectiveness of the programme has in some cases been compromised by either poor targeting or implementation in tandem with production-linked support policies that are associated with environmental problems (Brouwer and Lowe, 1998; OECD, 2003b). This is particularly relevant for horizontal programmes which are not oriented at a special environmental objective but aim at achieving low intensive management of land in general, as reported from Germany (Eckstein *et al.*, 2004). At the local level, there is evidence from Spain (Paniagua Mazorra, 2001) and Ireland (Emerson and Gillmor, 1999) that the voluntary aspect of participation has limited its effectiveness through the production of a patchwork effect.

5.3.3   AEP case studies

Boxes 5.1 to 5.3 summarise features of AEP schemes in three countries, i.e. Norway, Ireland and Spain. In Austria, the rather low level of intensive farming in large parts of the country has resulted in a very high overall participation in the horizontal agri-environmental programme (ÖPUL) since its establishment in 1995, with 72% of farm holdings participating in the scheme (farming 88% of the utilised agricultural area of Austria). The approach to the programme in Austria is characterised by an integrated approach that encompasses all agricultural activity (i.e. not just agriculture in ecologically sensitive areas as in some other countries). The aim is the "ecologicalisation" of Austrian agriculture covering the whole territory, and organic farming is an important element. Two types of ecological aim are pursued through ÖPUL:

- Maintaining the positive ecological effects of extensive and ecologically sound farming systems for the protection of biodiversity and landscape quality.
- Reducing the negative ecological effects of intensive farming systems by reducing degradation and soil erosion, contamination of ground- and surface water, and decreased biodiversity and landscape quality.

While there is a very high overall uptake of ÖPUL among Austrian farmers, participation rates among more intensive farms are lower, because they require more significant changes in existing farming management (Groier and Hofer, 2002). Moreover, the importance of alternative enterprises and diversification of the rural economy is borne out by the Austrian evidence (Groier, 2004). While the take-up of the programme is generally very high in mountainous regions, it is relatively less significant in areas with intensive winter tourism activity and where dynamic structural change is higher than average. Nevertheless, the extent of uptake and the quality of ÖPUL measures adopted by farmers has ensured a high achievement (Groier, 2004).

In Germany, underdeveloped remote rural areas benefit from protection measures for ground and surface water and reforestation schemes. The synergy of subsidies, funds, agri-environment measures and forestry measures contributes to maintenance of farming in poor farm regions. Agri-environment measures have much higher relevance for extensive farming areas, and much less for intensive farming because the financial incentive is not high enough. Agri-environment measures maintain farms producing in marginal areas but conditions of payment restrict land use intensity and therefore production.

Box 5.1: Agri-environmental Schemes in Norway

Norway has had environmental support or payment schemes in operation within the agricultural sector for many years. Since 2003, each farmer in Norway has had to set up an environmental plan for his/her farm to be eligible for any type of support from the government. In the future, the environmental plans for the individual farms will have to be in line with the agricultural environmental programme to be implemented in the county.

The first overall national programme was implemented in May 2004. The aim of the schemes has been to support farmers for the provision of public goods such as the cultural landscape, and to reduce negative externalities such as nutrients runoff.

In 2003, the two farmer unions and the government agreed to introduce regional environmental programmes from 2005, in addition to the national programme. From 2005, it will be up to the regional agricultural authorities and the County Governor to prepare regional environmental plans in cooperation with the farmers' unions and to establish the appropriate type of support schemes in each county. The regional environmental plan in each county will have to be approved by the Norwegian Agricultural Authority; the environmental support schemes must follow specific guidelines. A support scheme should either focus on the maintenance of the agricultural landscape (cultural landscape) or focus on reducing pollution from agriculture in the county.

One major environmental scheme in Norway is the acreage and cultural landscape scheme, which in principle covers all agricultural land in the country. The aim of the scheme is to keep agricultural land in use and thereby maintain an "open" landscape (agricultural land is scarce in Norway, only 3% of the total land area). In addition, the farmers have to obey specific rules regarding agricultural practices to preserve and take care of the totality of the agricultural landscapes respecting environmental values and recreational values, to be eligible for these payments ("cross-compliance").

Farmers in Norway also receive headage payments, partly as income support, and partly as environmental payments to enhance grazing or an environmental and animal welfare friendly husbandry in general. Organic farmers are entitled to additional acreage payments. The territorial effects of these "broad" schemes are difficult to judge since it is not known what would have happened without these schemes. However, it is reasonable to suppose that agricultural activity and land use would have been lower, especially in marginal areas. There is a concern in Norway that the areas of "open landscape" are diminishing, and that this may also reduce biodiversity since many plant and animal species are dependent upon these landscape types. Without these payment schemes, it is highly likely that the problem of overgrowth would be much larger than it actually is.

Box 5.2: The Rural Environment Protection Scheme (REPS) in Ireland

After 1973 (Ireland's accession to the EEC), agricultural modernisation produced economic and social benefits but negative environmental impacts, e.g.:

- pollution from silage effluent and animal slurry
- eutrophication of lakes and rivers, and contamination of ground water, from excessive fertiliser applications
- destruction of wildlife habitats due to land reclamation and drainage
- loss of sites of historical and scientific interest
- visual intrusion of farm buildings on landscapes
- increased emissions of methane from higher livestock numbers.

Despite these problems, there was very little agri-environment support in Ireland prior to 1992. However, the REPS was devised by the Department of Agriculture and Food and launched in June 1994, with objectives:

- to establish farming practices and production methods reflecting increasing concern for conservation, landscape protection and wider environmental problems/issues

- to protect wildlife habitats and endangered species of flora and fauna
- to produce quality food in an extensive and environmentally friendly manner.

Eligible farmers were entitled to a payment of €151 per ha up to a maximum of €6040 (1994).

The REPS comprised 11 horizontal measures, and 6 supplementary zonal measures, with the following characteristics: universal availability, voluntary, restriction of payments to under 40 ha, and the inclusion of a training element. The initial target of approximately 45,000 farmers (25% of total) and 1.3 million hectares was not met; by October 2003, there were 37,000 participants (29% of the total), covering 1,312,200 ha. The highest participation rates were principally in areas with small farms and extensive farming systems, while low rates were observed in the most intensive farming (and most damaged) areas.

From the various studies carried out in Ireland, the following conclusions may be reached:

- The REPS has been of greatest benefit to low-income small farms in more marginal farming regions (Department of Agriculture and Food, 1999; Matthews, 2002).
- Compensation payments have not been sufficient to encourage intensive farmers to adopt.
- Improvements in farming practices have led to reduced environmental impact (reduced application of inorganic nitrates and phosphates), but very little evidence of environmental enhancement (especially in relation to habitats and biodiversity).

There are particular concerns about lack of monitoring and the absence of specified targets.

Box 5.3: Three Agri-environmental Schemes in Spain

In Spain, there are regional or zonal programmes with two action areas, national parks and environmentally sensitive areas, and areas of specific environmental interest proposed by Regional Governments. The principal focus is the protection of low-intensity farming systems. However, because of the regional nature of the programme, the result has been a patchwork effect with limited connection among different regional measures, thus limiting its efficiency.

Case study 1: Habitat conservation: the Steppe Cereal Programme, Castilla y Leon

This programme aimed at the introduction of agricultural practices compatible with the conservation of steppe birds. The first two contracts were managed by the agricultural administration, and the last two by environmental authorities with a very different (i.e. strictly ecological, not rural) approach. The majority of those involved are part-time farmers.

Overall, the programme has succeeded in improving habitat quality, changing the homogenous landscape structure, and achieving favourable conditions for conservation. Farmers' participation depends mainly on economic factors, but geographical and socio-economic factors have contributed to an imbalance in the geographical distribution and so on territorial impact.

Case study 2: Unsustainable water extraction for agricultural production, Castilla la Mancha
Since the 1980s, a dramatic increase in crop production occurred because of an intensive programme of irrigation. The CAP had encouraged intensification and reinforced expansion of irrigation, with short-term improvements in employment and incomes but also environmental damage with overexploitation of aquifers, degradation of wetlands and loss of biodiversity. The resulting unsustainable imbalance between water demand and water supply led to a sharp decline in groundwater levels, and laws were passed from 1987 onwards to limit the sinking of new wells.

The application of an agri-environment programme during 1993–2002 allowed income compensation for reduction in water extractions. However, in spite of decreasing water consumption, the aquifers have not yet recovered (severe droughts in 1986/88, and in 1990/95).

Case study 3: Integrated rice production in salt marshes, the Guadalquivir River
This area is a very significant habitat for aquatic birds but is also a productive rice-growing region, and damage was occurring from pesticides and nitrogen applications. The introduction of integrated rice production systems in almost the whole of the rice area has resulted in reduced use of these chemicals.

In Hungary, an agri-environment policy was established 1999 to conform with EU regulations. General schemes are applicable throughout the country and specific schemes applicable in particular circumstances in selected zones. The National Rural Development Programme 2004–2006 comprises:

- entry-level schemes
- integrated crop management
- organic farming
- high nature value area schemes
- supplementary agri-environmental measures.

In total, over 15% of the utilised agricultural area of Hungary should be involved in agri-environmental measures by 2006. Take-up is biased towards less-favoured areas with higher unemployment and high dependence on agriculture. While agricultural policies are administered by the Ministry for Agriculture, compensation schemes for environmentally sensitive areas fall within the ambit of the Ministry of Environment.

5.3.4 Conclusions

The EU's Agri-Environment Programme is a response to the obligation to take account of environmental protection requirements arising from the Maastricht and subsequent EU Treaties. The diversity of the European agricultural landscape, and the diversity of cultural values and the differing structures of farming systems, make it very difficult to identify a common set of indicators to assess the effectiveness of the measures.

The objective of the analysis in this section has been to examine the overall impact of the agri-environment programme within the CAP as measured against ESDP and Cohesion objectives. While it is not possible to state with certainty what has been cause and effect in respect of particular policies, it is evident that the programme has the potential to contribute to the achievement of a number of the core objectives of the ESDP and ESPON. These can be summarised as follows:

- The Programme contributes to prudent management of and protection of nature and cultural heritage through encouraging a reduction in inputs of inorganic fertilisers, conservation of habitats, and preservation of the cultural landscape. Agri-environment schemes are particularly suited to the encouragement of appropriate land management (Baldock *et al.*, 2002).
- The provision of support for organic production, which is given a high priority in a number of countries, has the potential to contribute to balanced competitiveness through high-quality food production targeted at niche markets.
- The Programme makes an important indirect contribution to economic and social cohesion through the provision of income support in marginal areas and thus contributing to the retention of rural population.

While horizontal measures, especially in respect of organic production and training, have been a feature of the programme in most Member States, it has been largely identified with environmentally sensitive and extensive farming areas, with the notable exception of Austria, where the aim is the "ecologicalisation" of all agricultural activity.

5.4. Early Retirement Scheme for farmers

5.4.1 Introduction

The Early Retirement Scheme (ERS) aims to address the perennial structural problems of the elderly farmers and poor holding viability, main features in a number of Member States. It provides a pension for elderly farmers to retire and an opportunity for young farmers to take over and enlarge their holdings. Current eligibility criteria (EU Reg. 1257/99) include limitations in relation to age (the transferor should be between 55 years old and normal retirement age, whilst the transferee should not exceed a maximum age), occupation (the transferor must have practised farming for the preceding ten years, and must cease all commercial farming activity; the transferee must practise farming on the holding for not less than five years), economic viability (measured in terms of an obligatory increase in the size of the transferee's agricultural holding) and farming skills (the transferee having a Certificate of Farming or other adequate farming experience).

The Scheme is not mandatory, and hence has not been implemented in some countries, including Italy, Luxembourg, Sweden and the UK. It "has proven to be most popular in France, Ireland and latterly Greece, and these countries in aggregate accounted for 88% of total spending between 1992 and 1999" (Caskie *et al.*, 2002, p. 12).

5.4.2   Country applications

In France, an early retirement premium to the general state old-age pension for full-time farmers over 65 years old (Indemnité Viagére de Départ – IVD) was made available from 1968 to 1974 in problem areas such as Brittany and the Massif Central, and was associated with farm enlargement objectives, or "the installation of suitably qualified young farmers" (Naylor, 1982, p. 28). To this extent, the IVD constituted the institutional template on which the first mainstream EU ERS was built in 1992 (Regulation 2079/92), and which attempted to strike a balance between social and structural objectives.

In Greece, a first version of an ERS was implemented under Regulation 1096/88, and a second version under Regulation 2079/92. In Ireland, a first initiative came into operation between 1974 and 1985 under Directive 72/160/EEC, but only 600 farmers participated, "much less than an exploration of the potential attitudes towards the scheme of elderly farmers in the west might have suggested" (Gillmor, 1999, p. 80), and a further round took place in 1994. In Spain, the initial European package (Reg. 1096/88 and 3808/89) was linked to the restructuring objectives of young farmers and attracted very small numbers of beneficiaries (868 in 1990 and 1991) (Paniagua Mazorra, 2000, p. 115). The eligibility criteria became slightly more relaxed during 1993-1997 in the first round of the Spanish version of the ERS (Reg. 2079/92). In Finland, 5,569 ERS participants were reported in 1995–99 (Ministry of Agriculture, 2001). In Norway, a national ERS was introduced in 1999, with the important proviso that the transferee could not be the partner of the transferor.

The implementation of the EU ERS has been examined at a regional level in France, Greece, Spain, Ireland, Finland and Norway, with a distinct spatial pattern of adoption gradually coming into view. In France, the highest levels of adoption of the IVD grant (1963–78) were reported (Naylor, 1982) to be in areas of least need, and that "government support, through the CAP, for the maintenance of agricultural prices at levels which encourage small farmers to remain in business also conflicts with retirement policy". Moreover, areas of part-time farming appeared to have had lower than average levels of IVD adoption.

For the period 1992–94, Allaire and Daucé (1996) noted a rather strong regional contrast in ERS participation rates between the Paris basin and the littoral Mediterranean areas (around 15% of the eligible population) and the geographical crescent that includes Bretagne, Burgundy and Lorraine passing from the heart of the Massif Central (30%). Dairy regions most affected by restructuring showed the highest ERS take-up rates, whilst cereal and intensive cropping regions remained relatively indifferent. The adoption rate was higher for farmers with lower incomes (21% on average; Brangeon *et al.*, 1996). Regional discrepancies were also found in the rates of farm transfers to young farmers; these were largely explained by differences in average regional agricultural incomes and/or positive attitudes to the relevant institutions (Daucé *et al.*, 1999).

In Greece, 2,500 farmers (out of 8,151; Census 1991) participated in the first version of the ERS on the island of Lesvos, part of the North Aegean Region of particular disadvantage. However, this involved extensive fraud, and anecdotal evidence suggested that many farmers accepted participation without being aware of the accompanying regulations which resulted in their losing price subsidies (Koutsomiti, 2000, p. 54). This created considerable hesitation on the part of the Lesvian farmers towards later versions of the ERS. More than half of the transfer coming from the olive-growing areas of the island, while the tourist areas of the island, and those with a high pasture share, were hardly represented. The greater part of the agricultural land transferred under the ERS was planted with trees, especially olives. Comparison of the average holding size transferred with that received showed that the ERS contributed to a 72.5% increase in the size of holdings. However, nearly all were small-sized farms (under 5 ha), showing the extent of land fragmentation in Lesvos.

An important feature of the Lesvos case study was that 70.8% (transferors) and 62.5% (transferees) of the ERS-participating farmers were women (Koutsomiti, 2000). Men were more involved in off-farm employment and were thus not ERS-eligible. This is in striking contrast with the census data (1991) where 70% of the owners of agricultural holdings in Lesvos were reported as male (Koutsomiti, 2000). Moreover, 72.9% of the land transfers in Lesvos were intergenerational (from parents to children). Clearly, the ERS was used as a farm household strategy to increase income as a whole by "bending" regulations (or misreporting in census returns), maximising pluriactivity and "juggling" resources.

In Ireland, ERS applicants were geographically distributed in a highly unbalanced fashion (Murphy, 1997), with the majority of applicants coming from the traditionally more prosperous farming areas (Leinster, Munster, Connacht and Ulster). More than half of farmland transferred was in the dairying south-west counties, characterised by medium-sized to large farms, strong commercial orientation towards farming, and young Irishmen prepared to farm on a full-time basis, while "the lowest rates of participation in the ERS were in the west and north-west Irish regions, which are characterised by a higher proportion of unmarried farmers and small-sized, poorer, dry stock farms associated mostly with low-income cattle and sheep grazing activities" (Gillmor, 1999). Murphy

(1997) has pointed out the presence in the west of Ireland of many part-time farmers alongside significant numbers of elderly farmers who are single and without a readily identifiable successor. These elderly farmers were either too attached to farming and disinterested in retirement and occupational role changes (Murphy, 1997), or too old to qualify for the ERS and did not have spouses who would be able to qualify on the basis of joint management. Moreover, the small size of part-time farms prevailing in the west did not allow their owners to claim that they practised farming as their main occupation.

In Spain, the geographical distribution of ERS farmers (1990–94) was also concentrated, with regions of intensive farming and higher than average numbers of young farmers (e.g. Castille and León) attracting the majority of aid granted throughout the country (Paniagua Mazorra, 2000). However, "the farmers attracted to the retirement programme have holdings of insufficient economic size (average: 14.9 ha)" (Paniagua Mazorra, 2000, p. 116). The great majority of the ERS holdings were small-sized (half were less than 10 ha), owned (75.8% against a regional average of 59%) and fragmented (each holding is made up of 12.7 plots on average) in Castille and León. Most showed "very little intensive farming, although many are situated in high-yield districts" (Paniagua Mazorra, 2000, p. 118). The great majority of early retirees were married (74.2%), whilst only 19.8% were single and 5.8% were divorced.

In Finland, the territorial impact of the ERS has taken the form of a North-South divide (Pietola *et al.*, 2003). Farmers located in Northern and Central Finland were more likely than those in the South to retire early and transfer their farm to a new entrant. In Norway, a spatial pattern in ERS uptake was also displayed during 1999-2003 (Statens Landbruksforvaltning, 2003a). The NUTS3 regions with a higher uptake than the national average occurred in "strong" agricultural counties (e.g. Sør-Trøndelag and Nord-Trøndelag in the middle of the country, Rogaland in the southwest, and Oppland in inland Østlandet) with a higher than average employment in agriculture, animal husbandry and farm size. By contrast, the regions in the central part of South Eastern Norway (in the Oslo region), characterised by cereal production, off-farm employment opportunities and part-time farming, had a lower than average uptake. The very low uptake in Vest-Agder and Telemark in Southern Norway is harder to explain. Small-sized farms, part-time farmers and a less vibrant labour market prevail in these regions, with the low uptake being more associated with the traditional agricultural communities.

5.4.3   Conclusions

As Blanc and Perrier-Cornet also point out (1993, p. 322), inheritance practices "cover well-defined geographical areas that rarely correspond to national units", and geographical diversity in patterns of intergenerational farm transfers has been well reported at the European level (Lamaison, 1988; Perrier-Cornet *et al.*, 1991; Gasson and Errington, 1993; Tracy, 1997; Errington and Lobley, 2002). Comparative analysis of French and Illinois grain farmers (Rogers and Salamon,

1983) indicated that early retirement was the preferred farm exit option in communities favouring multiple-heir systems, regardless of their ethnic origins. Divisive inheritance was seen as responsible for low celibacy rates, small family sizes and low out-migration amongst the community members. In general, equal shares did not appear to provide the multiple heirs with employment and viable holdings but rather helped them to avoid permanent out-migration and retain social relationships with the farming community as valuable land-owning members. In this way, multi-heir systems encourage geographic immobility. Nevertheless, inheritance strategies (i.e. legal transfer of rights) vary significantly from country to country or even from region to region.

There have been numerous attempts to summarise inheritance strategies (Perrier-Cornet *et al.*, 1991; Blanc and Perrier-Cornet, 1993; Ross Gordon Consultants, 2000). In particular, "the combination of share-out in kind (possible and actual) and egalitarian practice" is common in many Mediterranean regions and "fosters the development of pluriactivity, retirement farming and even survival agriculture for the unemployed" (Blanc and Perrier-Cornet, 1993, p. 322–323). However, diversity remains paramount with certain Greek Aegean islands (e.g. Karpathos) even preserving the ancient "matrilineal" system of inheritance whereby the mother's property is passed down to her oldest daughter. "In Italy, primogeniture is still followed in the autonomous Tyrol province of Alto Adige" (Gasson and Errington, 1993, p. 196).

Regions in the Netherlands and Germany are characterised by inheritance strategies based on the need to preserve the unity of the holding and thus favouring unequal shares and full-time employment. However, the concept of splitting farms equally among all heirs (*Realteilung*) has been the prevailing inheritance system in the southwest and some areas of North Germany (Ross Gordon Consultants, 2000). In the UK and Ireland, single and non-compensatory systems are predominantly encountered. UK studies emphasise the impact of farm family cycle and the presence of a successor on farmers' decision making and land use (Potter and Lobley, 1992, 1996; Gasson and Errington, 1993; Wallace and Moss, 2002). A similar inheritance pattern of "keeping the name on the land" is found in Finland (Abrahams, 1991). The French inheritance tradition stands in the middle, with the farm successor paying "compensatory sums to the co-heirs, but the land is under-valued on average by half compared with open market prices" (Blanc and Perrier-Cornet, 1993, p. 324). Danish and Belgian regions follow the French inheritance pattern of equal shares and a single successor with parents rather than co-heirs being the transferors.

If regional demographic patterns tend to correspond to the inheritance strategies preferred, as Rogers and Salamon (1983) claim, then the ERS's differentiated territorial impact may be more correlated to demographic indicators than to farm succession issues and the farm family cycle itself. Such correlation will be examined in the following section. The regional imbalances identified in the ERS at the EU level and its higher than average uptake in dairying farming regions can be hypothesised to be more associated with demographic matters, social organisation and the absence of young successors rather than sectoral features *per se*. Statistical analysis would be required to substantiate this claim across all relevant regions at the EU level. Fennell (1981) has argued that there is a correlation between the level of urbanisation in a region and farmer retirement: "basically retirement is an urban concept and farmers in some regions and countries are more immediately affected by urban values".

Comparative regional analysis at the NUTS3 level was carried out amongst the countries with the highest ERS rates, using two contrasting regions with minimum and maximum uptake levels. This analysis suggested that ERS uptake increases proportionally to population density. Further statistical investigation showed that densely populated, leading, meso-accessible and non-LFA regions attract the highest numbers of ERS participants.

It has been questioned whether the EU's ERS is beneficial or detrimental on environmental grounds. Only a few participating farmers have exercised the ERS option of reassigning agricultural land to non-agricultural uses when it cannot be farmed under satisfactory conditions of economic viability. As Gillmor explains for the Irish case (1999, p. 85), "while transfer of land to non-agricultural transferees and reallocation of land to non-agricultural uses, forestry and ecological reserve creation were permitted under the ERS (1994–99), there has been little use of this allowance". The existence of alternative income sources from both the Rural Environment Protection Scheme and the Afforestation Scheme were found to lessen the likelihood of some farmers participating in the ERS, which has contrary objectives (Gillmor, 1999).

To this extent, the ERS embraces environmental contradictions and wider policy dilemmas in its objectives. One view is that "a change in farm occupancy, which leads to amalgamation of farms, 'is one of the major factors in landscape change from agricultural intensification'" (Caraveli, 1997, p. 172), and that "by transferring land ownership and management to younger full-time farmers who are likely to work the land more intensively and by promoting farm enlargement, the ERS is more in accordance with the principles of productivist agriculture" (Gillmor, 1999, p. 85). In the UK context, young farmers (recent farm successors) have been associated with dramatic land use change, intensification and, consequently, greater environmental impact (Potter and Lobley, 1996). "According to another view, though, larger farm sizes could 'create conditions compatible with extensive production systems, as small and fragmented farms cannot easily adopt extensive production practices'" (Caraveli, 1997, p. 172).

To conclude, there are some important points that can be made in relation to the highly differentiated territorial impact of the Early Retirement Scheme (ERS):

- A distinct spatial pattern of adoption of the ERS exists (France, Ireland, Norway, Finland and Spain): the highest levels of adoption were reported in areas of least need (i.e. prosperous farming regions) and amongst higher numbers of young farmers. Population density emerges as an indicator of the regional propensity to early retirement.
- In countries with the highest rates of participation (France, Greece and Ireland), the structural effect was little different from that which would have occurred anyway, albeit over a slightly longer time scale (it did not increase

the rate of retirement in the long run and did not encourage farm transfers outside the family).

- LFAs are characterised by higher than average sensitivity to the timing of exits from farming (Greece). The time gains offered by the ERS are important in relation to the depopulation problems and the demographic scarcity of farm successors prevailing in LFAs where the younger generation's rejection of farming (as a career) due to delays in farm transfers leads not only to alternative employment but also to out-migration.
- There is strong sectoral attraction for dairy/intensive farming regions (France, Ireland and Norway) and/or high-yield regions (Greece and Spain). Cereal regions remain largely indifferent to the ERS (France and Norway).
- A pension income higher than existing earnings emerged as the main explanatory variable for those who decided to participate in the ERS (France, Ireland).
- The ERS adoption rate is higher amongst farm households with lower than average incomes (France, Spain, Greece and Finland).
- The absence of a successor, and a farmer's single marital status, decrease the likelihood of exit from farming (UK, Finland, Spain and Ireland).
- The environmental impact of an ERS is highly dependent on the national context and perspective.

5.5. The Less-Favoured Areas scheme

5.5.1   Introduction

Although the LFA scheme (see Chapter 4.6) is the CAP instrument which addresses the territorial dimension of agricultural production most directly, its impact cannot be assessed only by the analysis of this single measure. The dominant objective of the LFA instrument is to maintain farm management in less-favoured areas based on environmental principles and provision of other functions beyond food production.

Land used for cattle, sheep, goats and dairying is generally eligible for LFA payments; in most countries, payments on cropped land are restricted or reduced (e.g. France, Austria, Germany). In Mediterranean countries, where cropping is widespread in the LFAs, some or all of the cropped area is typically eligible for payments. In many countries, allowance rates were increased under Regulation 1257/99 in order to ensure that there are few losers from the change to an area-based system. The increases also compensate for any additional costs associated with good farming practice.

Specification of the maximum level of compensation in Reg. 1257/99 does not enforce differentiation between types of LFAs, but the level of LFA subsidies varies considerably between different countries and regions (Dax and Hellegers, 2000; ECA, 2003), reflecting the priorities of the Member States, and the criteria and approach used. Criteria for allocation of payment include: age of farmer, type of area, type of cultivation, with:

- considerable divergence of average payments per hectare and holding, between "northern" (more) and "southern-med" (less) MS
- in the south, resources are focused more on modernisation schemes and improvement of processing and marketing structures
- small structures of farms in the south often beneath the limit of eligibility
- modulation or limitation of payments or farm size (ha) eligible for payment in some countries.

The range of differentiation between low-input farming systems and intensive upgrading farming (e.g. livestock numbers) is quite large between the Member States. Also, many of the new Member States prepared their classification systems before EU accession in 2004 and focused strongly on LFA support (e.g. Slovenia). This differentiation can be characterised by:

- a wide range of indicators (income, labour force input, types of farming etc.) which leads to disparities in treatment and application
- in order to differentiate application of instrument, farmers and/or areas are classified into "groups", "zones" and "scoring systems"
- a horizontal-geographic approach (e.g. Finland, France, Sweden) or a more vertical one (e.g. Austria, Germany, based on individual farm production levels)
- particular differentiatiation within LFAs in Austria, with a refined, detailed scoring system for mountain farms

Some of these features are addressed by the Commission's July 2004 proposal (EC, 2004a) for the 2007–2013 rural development programmes, aiming at a review of the classification of the intermediate zones and to lower the maximum payment of the intermediate zones to 150 €/ha.

Although the co-financing rates show considerably higher levels for Southern European countries, the uptake of compensatory allowances has been particularly weak there. The different implementation and use of the measure is reflected in the statistics of the uptake, showing marked differences between Member States. In some countries like Italy, Germany and Spain the regional administrations are responsible for the running of the scheme and adapt it to local circumstances. Thus national averages have to be differentiated for the regions and types of LFA (mountain areas and other LFA). Whereas the total of just over 1 million farm holdings in the EU-12 which benefited from the scheme in 1994 accounted for 45% of all holdings in eligible areas, the participation rate varied from 9% in Italy to between 84% and 99% in most northern Member States (CEC, 1997, p. 55).

The main reason for the lower proportion of farmers in the countries of the South receiving aid out of the total number of farmers in the LFAs is inherent to the concept around which it has been built (Terluin *et al.*, 1995). The orientation of the compensatory allowances scheme on headage payments made it obviously more applicable in regions which focus on livestock production, including Ireland and the UK, but also Greece. In particular, the small structure of farms in the South, with many farms of a size beneath the eligibility threshold, excluded a large proportion from payments, in spite of the fact that the minimum limits for the granting of aid in these countries has been lowered. Thus many farms are not eligible, e.g. in Italy, where 29% of farms are less than 1 hectare in size. Moreover, the difference in the levels of payments for livestock and crop production disfavour the application of the scheme in regions where permanent cultures and arable land have a significant proportion in land use. The difference is most outstanding between mountain areas in the North and the South: whereas in the North arable land and permanent cropping are of no relevance in mountain areas (and of limited relevance in other LFAs), they are a marked feature of land use in the Southern LFAs.

Another reason for the lower commitment of Southern Member States can be found in the focus of allocating funds on modernisation of holdings (EU Reg. 2328/91), improvement of processing and marketing structures (EU Reg. 866/90), and less on compensatory allowances. The different priorities identified by Member States and the great variety of policy implementation, including modulation of payments etc. lead to considerable differences in the uptake which are not to be explained by structural differences alone. Factors of importance, among others, include:

- Although the average payment per beneficiary holding showed a high variation between Member States in the 1990s, the divergence increased and it ranges now between €800 and €7,000. In the regions most concerned, LFA support achieves up to about 40% of farm income (CJC Consulting, 2003, p. 54).
- The same diversity in the uptake of the payments does not only affect the level of payments per beneficiary holding but also the proportion of beneficiaries with regard to all holdings in eligible areas. This proportion varies from 9% in Italy to nearly a total coverage of farmers in some northern Member States (Ireland, Netherlands, United Kingdom).
- The implementation of the scheme by Member States and regions greatly affect the uptake and budget spent for the measure: whereas some countries do not modulate the payment according to the size of the holding, in others provisions exist to differentiate grants according to type of production, number of productive units, stocking rate, maximum payments or revenue of the farmer.

5.5.2 LFA interaction with other CAP/RDP instruments

The Second Pillar includes a relatively small proportion of total CAP funds, but the decoupling process has opened agricultural policies to overall rural development and could facilitate turning some of the natural handicaps of mountains and other LFA into advantages: for instance, cultural heritage, landscape, high-quality products, diversification (Nordregio, 2004). As the maintenance of agricultural land use in these areas is more important than production, a number of other policy instruments are relevant in supporting these aims, including:

- agri-environmental programme
- other RD measures (investment, setting-up premiums etc.)
- market premiums and compensatory allowances (CAP regime)
- other systems of transfers to rural areas.

In a rising number of regions, the important role in maintaining multifunctional cultural landscapes is addressed explicitly in programmes including various instruments (see definition of "multifunctional agriculture" in Switzerland in 1996; rural amenity provision in mountain areas of Austria, OECD 1998b, and e.g. initiatives of Alpine Convention).

Other instruments may exercise an effect in both directions: positive as supplement and reinforcing activities or adverse effects (trade-off of objectives):

- counter-productive side-effect of the CAP premiums and compensation allowances – incentives for production (over-grazing) (Beaufoy *et al.*, 1994)
- sectoral/commodity instruments are not able to cope appropriately with the needs of LFA – mainstream CAP support is not oriented to extensive farming system (Dax and Hellegers, 2000, Swedish case study)
- low agricultural incomes and less developed regional economies in LFAs often go hand in hand, therefore cross-sectoral approaches are essential;
- lack of coordination with other systems of transfers to rural areas (Swedish case study).

5.5.3   Impacts

Initially, the prime focus of LFA policy was not on the impact of agriculture on the environment. The criteria for designating LFAs were intended to reflect the degree of disadvantage for agricultural production, not environmental value or problems (Beaufoy *et al.*, 1994). Nevertheless, there is a great deal of overlap of LFAs with regions of High Nature Value farming systems. Overall, the impacts of the LFA scheme on *land use* can be assessed as:

- Environmental impacts are relatively minor in the short term, with no stringent, conclusive evidence about the impact of LFA on the environment, only contextual interpretation (Dax and Hellegers, 2000).
- However, the environmental impact may be assumed to be substantive in the long term, e.g. maintenance of farming structures and land use, underpinned by analysis on regional trends of mountain farming over the 1990s (Baldock *et al.*, 1996; Dax, 2002b).
- In general, low-intensity farming systems are mostly situated within the LFAs; but this does not mean that these areas are automatically dominated by environmentally friendly cultivation (Beaufoy *et al.*, 1994).

- Farms within the LFA vary greatly in their conservation performance (Beaufoy *et al.*, 1994).
- The LFA scheme even may encourage extension of farming into fragile areas and valued habits, and provide incentives to maximize livestock numbers on a holding (overgrazing) (Baldock *et al.*, 1996; Buckwell *et al.*, 1997).

LFA scheme provides a substantial contribution to farm incomes. According to the differences in application described above, this effect varies considerably. Case studies have shown that it attains a significant level of more than 10% in many regions, including Austria with 19%, France 1–15% (for simple LFAs) and 22–38% (for mountain regions), and Finland 42% (CJC Consulting, 2003, p.54). Further aspects are:

- Even if the level of subsidies varies considerably between different regions within the community (Dax and Hellegers, 2000; ECA, 2003), it contributes significantly to the income of low-intensity farming in many areas (Beaufoy *et al.*, 1994);
- Other social transfers and benefits which are used by the agricultural holdings must also be taken into account (e.g. Sweden);
- The contribution of compensatory allowances to farm income has increased considerably over recent years in some countries with major LFAs (e.g. Austria, France). This has helped to decrease the income gap between mountain farmers and non-LFA farmers in some situations (Dax, 2004). In some Member States (Portugal, Spain, Greece and Italy), it has only a modest contribution to the income of farm households.

It is often argued that *out-migration* would have been higher without support schemes like LFA. The impact on amenity provision and landscape development has an effect on the overall regional economic activities, and, particularly in areas with high tourist potential, is highly relevant for regional performance (Dax, 2004). A recent study on the European mountain areas (Nordregio, 2004) reveals that different processes of demographic change are taking place; the general trend is that depopulation in mountain areas is higher than in lowlands. Yet in north and central Europe there is a stable or even positive population development, whereas in eastern Europe depopulation is the norm (Nordregio, 2004, p.c.). However, the direct impact of agricultural policies on these trends seems to be limited. Other case studies (e.g. from Sweden and Austria) support that out-migration is less rapid or similar for LFAs compared with non-LFAs.

5.5.4   Conclusions

The spatial differences of European agriculture are reflected in the application of the LFA scheme. In contrast to what one would expect from a compensation measure, the application of the scheme is largely correlated to the degree of farm net value added, i.e. higher Compensatory Allowances are applied in more prosperous countries, and in "poorer" countries only a low level of Compensatory

Allowance is achieved. The lower commitment of southern Member States is partly due to the prevalence of arable land and permanent cropping in the LFA of the South (whereas the scheme is largely oriented towards livestock farming) and the focus on modernisation schemes and the improvement of processing and marketing structures (Dax and Hellegers, 2000, p.184ff; Shucksmith *et al.*, 2004, Map 4.8, p.94).

A major reason for the spatial distribution of funds is the reference level, which is set at the national level and not at the European level, which implies that differences between Member States remain unchanged.

The steady extension of the LFA area over the decades of application reflects the difficulty of adjusting the border of LFAs, and gives rise to further discussion on the criteria of delimitation and internal differentiation. The review of the intermediate zones as proposed by the Commission in July 2004 will address this issue. As the extension has been partly accompanied by an increase of overall grants, at least in some countries, the support level per unit did not reduce.

- The recent changes of the LFA scheme had an impact not only on the farm management itself but also on the farm income. In several countries the changes were cushioned by an increase of CA funds and thus had a positive impact on farm income in LFA. At least for several countries, this effect can be analysed (Austria, Dax, 2004; Hovorka, 2004; Finland, France, Germany, Spain, etc., CJC Consulting, 2003).
- There is a strong linkage to high nature value (HNV) farming systems and overlap is quite marked. The existence of HNV farming systems in these areas points to their beneficial role for nature conservation and biodiversity. These farming patterns are, however, highly threatened by impending marginalisation processes, which is particularly relevant for peripheral situations, including regions of the new Member States.

5.6. The LEADER community initiative

5.6.1   Introduction

The LEADER programme, started in 1991, is the EU scheme designed for the development of rural areas, and designed to help rural actors improve the long-term potential of their local region via Local Action Groups (LAGs). Its approach looks for innovative strategies for development of rural areas. The core elements of the programme are the preference towards integrated regional development strategies against sector-specific measures, the requirement to focus on the participation of local population and the intensive cooperation and networking in rural development activities.

The LEADER programme is now in its third generation. LEADER I marked the beginning of a new approach to rural development policy in 1991, which is territorially based, integrated and participative. In many aspects, LEADER I was a pilot scheme which led to a "reconsideration of traditional delivery systems for rural development support" (Dethier *et al.*, 1999, p. 179) at national and regional levels. In LEADER II (1994–99), this approach experienced a considerable expansion, with an emphasis on the innovative aspects of projects. In that period, the number of LAGs rose substantially and implementation affected a number of areas almost five times greater than in the first period. LEADER+ (2000–2006) continues its role as a laboratory for the emergence and testing of new methods of integrated and sustainable development, combining an endogenous approach with an approach of cooperation, networking and mobilisation. It has a strong focus on partnership and networks of exchange of experience.

A number of cases have been analysed in this project in more depth to highlight the impacts and linkages to CAP. This will be particularly important in advancing the possible "mainstreaming" of the Initiative in the forthcoming 2007–2013 period.

5.6.2   The LEADER method

LEADER is based on seven major components which are briefly outlined below. The combined application of these LEADER features is referred to as the *"LEADER method"*, which concentrates on local, trans-local and vertical features. Differences to "mainstream" Structural Funds programmes are conceptualized as follows (ÖIR, 2003):

- *Area-based approach:* Development is focused on a specific territory. The better use of endogenous resources, the horizontal integration of local activities, the strengthening of common identities and a shared vision for the area are key issues of an area-based approach.
- *Bottom-up approach:* The active participation of all interested people and organisations in planning, decision making and implementation of social and economic development is encouraged. More clearly identified local problems and needs, a better organisation and a greater acceptance of local decisions at various levels are the main advantages of this approach.
- *Partnership approach:* The engines of local development are the Local Action Groups (LAGs) within which rural stakeholders (individual persons or collective bodies – based on a contract binding all partners under the same conditions and for the same purpose) design rural development measures, at local level, that best suit their requirements, and develop and implement common strategies and innovative measures.
- *Innovation:* The main aim is to give new answers to existing problems of rural development, which provide added value and increased territorial competitiveness.
- *Multi-sectoral integration:* The approach contains both the combination of activities of different economic sectors and public and private activities in one project, and the strategic coherence between different projects in respect of a common vision.

- *Networking and transnational cooperation*: The capacity and readiness for collective action, to work for a common purpose within LEADER groups and other independent actors and cross-border cooperation between LEADER groups located in different Member States, is viewed as an important source for a common understanding and development of rural Europe.
- *Decentralised management and financing*: Apart from Operational Programmes, Member States are free to choose the intervention mode within the "global grant", which is characterised by the transfer of the budget for the local action plan to the local partnership. The local group is entitled to allocate the funds to project promoters according to rules set by the national or regional programme administration.

5.6.3   Impact of LEADER

- The wide application of the LEADER approach has had an impact on many rural regions of EU-15. Other countries, including new Member States, have adopted the programme philosophy, and have created similar initiatives adapted to their specific contexts. Indeed, the horizontal application of LEADER since the second programming period (1994–99) has led to a "race" of regions to be included in this scheme.
- One of the prime effects was the impact on the quality of the regional development process. The approach turned the attention to enhancing local partnerships and focusing on the endogenous local/regional development. With on-going experiences there have been adaptations to the strict orientation on small-scale issues, enlarging the regional development considerations to issues of transregional cooperation and integration of economic development into the larger spatial economic tendencies.
- The effectiveness of the initiatives is largely dependent on the institutional framework of the region, and its understanding of its role and development potential. This has been described as the "institutional thickness". Local/regional partners and institutions mostly have to undergo a long-term process to achieve substantial effect which is greatly reliant on the level and type of cooperation, and on many items summarised as social capital available in the regions. LEADER has raised awareness of these intangible factors of rural development, and provided a forum to prepare and enhance rural development strategies and initiatives. The actual impacts are very context-specific, to be expected as the outcome from a highly localised programme, being applied as a type of pilot scheme seeking for innovative processes and combinations of activities for rural development.
- In most cases, participation could be raised substantially within the regions. This has also been communicated as one of the particular positive outcomes to other regions and people from outside. The detailed issues and

commitment of regions is affected by national influences and the support being provided by the network structures for the LEADER initiative.

- With regard to the geographical distribution of projects within LAGs, there is evidence from the Irish LEADER II evaluation that the geographical distribution in most regions is uneven. There are tendencies towards local clustering in quite a few regions, which points to the pivotal role that towns and villages have in the implementation of local area-based approaches to rural development. In other regions, more dispersed patterns are evident, but it would seem that this has only been achieved in those areas where a deliberate strategy of spatial targeting was adopted.

- The development of LAGs in Austria shows that in the LEADER II period the LAGs were situated within or adjacent to the mountain areas with a population density far below the Austrian average. Comparing LEADER+ with LEADER II projects reveals a considerable extension of LAGs across Austria (from 31 to 56 LAGs). The LAGs, which are still located in the more peripheral regions, have grown in number and extent towards the main towns. This development may lead to the situation where the influence of, and concentration towards, cities will grow but at the same time may provide a chance to build up and strengthen the relationship between urban and rural areas.

- LEADER activities contributed to the sustainability of development processes at the local level. In Austria many LAGs already constituted under LEADER II are again part of LEADER+ and also products and instruments acquired and developed under LEADER II are still available (e.g. Cheese Route Bregenzerwald). In other cases where partnerships have ceased their activities within the programme the importance of local partnership is still tangible and many new partnerships, local development agencies and cooperation structures have sprung up and contributed to the diversification and dynamism of rural territories. LEADER thus has provided a particularly important phase of institution building for the regions (Koutsouris, 2003).

5.6.4   Conclusions

Ex-post evaluation of LEADER II summarises the programme both as efficient and effective. It proved to be *adaptable* to the different socio-economic and governance *contexts* and applicable to the small-scaled area-based activities of rural areas. It could therefore also reach lagging regions and vulnerable rural territories. LEADER activities induced and conveyed responsibility to local partnership, linking public and private institutions, as well as different interests of various local actors, to a common strategy. A profound change from a passive to an active attitude could be achieved among many local actors. In countries with a long-standing tradition of pluriactivity, agricultural diversification served as a basic pattern for multi-sectorial strategies, often in combination with rural tourism. A good example for the multi-sectoral approach based on agricultural products and rural tourism is analysed in the Austrian LEADER case study.

In a series of Member States such as Germany, many of the LEADER projects focused mainly on environmental measures trying to protect and further develop existing natural capital. The building up of partnerships and common regional activities like "Nature Park Uckermärkische Seen" or projects ranging from regional marketing, renewable energy or agricultural pilot projects were bound to maintain or develop the sustainable, and environmentally friendly use and exploitation of the natural capital. Moreover these activities have been supplemented in some countries (e.g. Germany, Spain) by national programmes, which underlines the need for regional programmes of this type within rural regions.

LEADER and its approach has some specific features, summarised in the term "LEADER method", which may lead, despite a limited budget, to specific outcomes and regional effects. Measures financed by LEADER projects are of a smaller scale and of a more experimental character than other Structural Funds instruments, and they provide a broader range of beneficiaries, especially from the non-profit sector, and female entrepreneurs.

Direct positive effects on employment cannot be easily quantified. An estimation (of the evaluation study) suggests that up to 100,000 permanent full-time jobs have been created or safeguarded in the course of LEADER II. More income has been generated by new employment, more visitors and more value added from local products.

LEADER is not an instrument to change local economic structures or revalue local economy in an extensive way (BMLFUW, 2003). LEADER is an instrument to stimulate processes in the local economy rather than to promote investments. Many core projects do preliminary work in activating rural actors which is the background for further economic activities. The potential of LEADER lies especially in the improvement of intangible factors, in raising awareness, in strengthening strategy and cooperation within the region. This often builds the basis for the provision of better services and more competitive products.

The future integration of LEADER+ into the rural development programming (*mainstreaming*) as outlined in the Third Cohesion Report might have severe implications for the administration and contents of the LEADER activities. The specific features of the Community initiative should be maintained (and elaborated) in order to use the potential. LEADER II was very effective in creating new links between local actors and stakeholders, (re)building trust across contemporary social divides and sectoral points of view. However, the cooperating and the development of a common strategic planning need time and LEADER issues like multi-sectoral integration, networking and transnational cooperation between rural areas were often too ambitious for the LAGs (transnation cooperation) or were achieved only by the more advanced groups. The successful implementation of multi-sectoral integration was an effect of certain preconditions and external influences, e.g. a favourable administrative context, a thriving and diversified local economy, a viable, dynamic, representative mixed partnership and a strong strategic orientation in the local action plan, rather than of LEADER activities (ÖIR, 2003, p. 26).

5.7. The SAPARD programme

The Special Action for Pre-Accession measures for Agriculture and Rural Development (SAPARD) programme for the countries of Central and Eastern Europe (the New Member States, or NMSs) was briefly described in Chapters 2 and 3. Alongside the parallel (but larger) ISPA programme which focused on general transport and water infrastructure, SAPARD was designed to assist the adaptation of agricultural structures and policies towards those of a market-oriented economy and the CAP/RDP, in particular by supporting rural development via creating agencies capable of designing and operating programmes funded largely by the EU. This subsection briefly reviews the allocation and use of these funds in the NMSs from an agricultural and rural viewpoint. The following section contains a more detailed case study analysis of the SAPARD programme in Poland.

Table 5.2 shows the indicative annual allocations for the ten NMSs over the 2000–2006 programming period; in many cases, these amounts will not have been expended in the earlier years due to delays in agency approval.

Table 5.2: Allocations for SAPARD and ISPA programmes (indicative annual allocations, 2000–2006)

| Country | SAPARD | | ISPA | | |
	Amount in million Euro	Share (%)	Amount in million Euro (min.)	Amount in million Euro (max.)	Average share (%)
Bulgaria	52.124	10.02	83.2	124.8	10.00
Czech Republic	22.063	4.24	57.2	83.2	6.75
Estonia	12.137	2.33	20.8	36.4	2.75
Hungary	38.054	7.32	72.8	104.0	8.50
Lithuania	29.829	5.74	41.6	62.4	5.00
Latvia	21.848	4.20	36.4	57.2	4.50
Poland	168.683	32.44	312.0	384.8	33.50
Romania	150.636	28.97	208.0	270.4	23.00
Slovenia	6.337	1.22	10.4	20.8	1.00
Slovakia	18.289	3.52	36.4	57.2	4.50
Total	520.000	100.00	878.8	1201.2	100.00

Source : *AgraFood East Europe* no. 216, Sept. 2000.

SAPARD provides applicant countries with the possibility of funding projects in the areas presented in Table 5.3. Out of the wide range of measures, four were selected as priorities by all applicant countries: investments in agricultural holdings (20% of the total public aid in all 10 countries), processing and marketing (26%), agricultural diversification (11%) and technical assistance. Two measures were taken up by 6–7 countries: rural infrastructure (20%), and environmental protection and maintenance of the countryside (i.e. pilot agri-environment schemes). Of the 9 other measures in the programme, none averages more than 4% of the total public aid. Although the balance differs from programme to programme, in virtually all of the candidate countries the share of public aid accounted for by the three most used measures is over 60% of the total (Wilkinson and Korakas, 2001).

Table 5.3: Priorities for SAPARD support measures

Measures	Priority
Investments in agricultural holdings	XXX
Improving the processing and marketing of agricultural and fishery products	XXX
Improving the structures for quality, veterinary and plant-health controls, for the quality of foodstuffs and for consumer protection	X
Agricultural production methods designed to protect the environment and maintain the countryside	XX
Development and diversification of economic activities, providing for multiple activities and alternative income	XXX
Setting up farm relief and farm management services	
Setting up producer groups	X
Renovation and development of villages and the protection and conservation of the rural heritage	X
Land improvement and reparcelling	X
Establishment and updating of land registers	
Improvement of vocational training	X
Development and improvement of rural infrastructure	XX
Agricultural water resources management	
Forestry, including afforestation of agricultural areas, investments in forest holdings owned by private forest owners and processing and marketing of forestry products	X
Technical assistance for the measures covered by this Regulation, including studies to assist with the preparation and monitoring of the programme, information and publicity campaigns	XXX

Note: XXX = all countries; XX = 6–7 countries; X = few countries
Source: European Commission, Directorate General for Agriculture, 2000a; Cunder, 2004.

Other measures, such as support for producer groups, water resources management or forest measures, have only been taken up by some countries with a specific interest therein. Direct payments similar to the LFA scheme are (together with horizontal agri-environmental measures) not elements of the SAPARD programme. Although a number of pilot actions address the need for more integration of local populations into the planning and operation of agricultural and rural development schemes and for models designed specifically for the problems of peripheral areas, experiences are rather scattered and not led by a strategic approach. In recognising the difficulties of the first experiences with the involvement of local bodies, the financial agreements for 2002 aimed to strengthen the bottom-up approach (CEC, 2002).

Given the early implementation state of the SAPARD programmes in general, it is not yet possible to achieve a detailed evaluation of socio-economic and environmental impacts (see Dwyer *et al.*, 2002). Led by the conviction that the Single Market and the support system of CAP cannot function without harmonised standards and procedures, the EU programmes for support for pre-accession aid focus on facilitating adaptation of national legislation as well as the administrative structures and procedures of the EU *acquis*. However, this approach leaves little room for national priorities or local bottom-up initiatives. There is therefore strong criticism relating to the focus of SAPARD capacity-building, whose emphasis seems misplaced since many candidate countries have a background of strong central state structures but weak local and non-state structures.

However, when assessing the implementation of the SAPARD programmes, the Copenhagen Summit agreement that ten new member countries (including 8 NMSs) could join the EU on 1 May 2004 shortened the programme period of SAPARD for most NMSs, and laid down provisions for programmes of rural development measures to be established as soon as the countries are EU members, including conditions more favourable than those applied to the EU-15 Member States.

5.8. Country case study: Ireland

5.8.1   Introduction

This case study aims to illustrate economic adjustment strategies of farm households in Ireland, in the context of structural change in the farm and rural economy and having regard to major policy measures. These strategies are considered under three headings:

- adaptations within conventional farming
- diversification (but on-farm) from conventional patterns of production
- uptake of non-farm employment.

A fourth section reviews trends in employment and enterprise development in the regional economy, on the basis that the recent unprecedented growth in the Irish economy is a major factor influencing farm household adaptation strategies. The final part identifies the main lessons that have wider implications for agriculture and rural development policies.

The selection of Ireland for the analysis of the interplay of individual, local and regional factors was supported by the availability of region-specific information and experience on structural changes. Moreover, some figures on the Irish farm and rural economy may be helpful in understanding the context and analysis, which follow. Primary agriculture accounts for 2.7% of national GDP, 6% of employment and 4.6% of exports (2002). The corresponding figures for the agri-food sector (primary agriculture plus food and drinks) are 8.5%, 9.0% and 8.3%. Of the country's agricultural area of 4.4 million hectares, 80% is devoted to grass (silage, hay and pasture). By far the greater part of production comes from livestock and livestock products. Beef and milk production currently account for 56% of agricultural output at producer prices.

5.8.2   Adjustments in conventional farming

Rationalisation of production units through *scaling up* the size of farm businesses is a common mode of adjustment under modern farming conditions. For the period 1991–2000, the size of the average farm in Ireland increased from 26.0 to 31.4 hectares, while the scale of business per farm, measured in European size units (ESUs), grew from 11.6 to 20.7. Correspondingly, the number of farms declined by 17%, from 170,600 to 141,500. The 1991–2000 data continue a long-established trend of a declining number of farms with an increasing average farm size. When trends by size classes (in hectares) are compared, there is a clear contrast between the lower and higher ends of the scale. The number of farms under 20 hectares declined by 46% during 1991–2000, while those over 50 hectares increased by 23% (Crowley *et al.*, 2004).

Upscaling is related to the existing regional distribution of farm business size. A consequence of this is that the farm size polarisation between regions is accentuated. In 1991, the difference between the Objective 1 and Non Objective 1 regions in average ESUs per farm was 8.8 ESUs, but this had increased to 13.8 ESUs by 2000 despite higher rates of increase in the Objective 1 region. Regions differ in the degree of intensification in farming with clear gaps evident between those in the west/northwest and the other regions.

Traditionally, scale enlargement was accomplished through farm amalgamation, by purchase or by *renting*. During the 1990s, however, the amount of land coming on the market declined considerably while its price increased. By contrast, the amount of land rented-in expanded. Apart from the contraction of the land market, there were other factors responsible for this expansion. Early retirement pensions were payable to farmers on condition that, *inter alia*, they transferred their holdings by long-term lease to family (or non-family) members. There is also a rental income tax exemption for land leased out on a long-term basis which also encouraged more farmers to opt for land leasing arrangements. Between 1991 and 2000 the amount of land rented-in increased from 12.5% to

18.7% of the total stock of agricultural land. Nearly one-third of all farms rented-in some agricultural land in 2000, compared to just over one-fifth in 1991.

Generally, renting-in of agricultural land is related to the larger and more commercially oriented farms in the east and south. By contrast the lowest rates of renting-in are in the poorer land quality areas especially along the west coast (Border, West, West, Mid-West and Southwest). However, while a north-west to south-east gradient of low to high renting is evident, the pattern is complex. The areas with the highest percentages of farms renting land are where specialist tillage and dairying enterprises are most prominent.

The 1990s saw a continuation of the long-term trend towards the reduction of *labour* on farms and to a less labour-intensive agriculture. There is little evidence of regional variation in this trend but labour intensities are somewhat lower in the more commercial farming areas (Non Objective 1), where underemployment is less likely and mechanisation is more advanced. Decline is observable not only in the numbers of persons contributing labour but in the amount of labour supplied by the workers. Nationally, the number of persons declined by 17.5% between 1991 and 2000, with little variation by region. Annual work units in the same period fell by 32.3%, again with no major regional differences.

The decline of 17% in the number of farms nationally during 1991–2000 was not spread evenly over holdings with different *enterprises*. The largest percentage decline (36.7%) occurred among "specialist dairying" farms, especially the smaller farms in this category. This trend is of considerable economic significance as it is on these farms that the largest gross margins are achieved. By contrast, there was a slight increase (0.5%) in the number of "specialist beef production" farms, which includes farms with the smallest average gross margins. Specialist tillage holdings had a relatively low rate of decline and specialist sheep farms declined at approximately the national average figure. Regionally, the switch away from dairying was most pronounced in the Western and Border regions, which include those areas with the weakest tradition in dairy farming.

Ireland's agriculture is based predominantly on livestock farming, especially beef and milk production, but there are clear regional differences in the distribution of these enterprises. About 59.5% of farms in the Objective 1 region are engaged in specialist beef production, with only 10% in dairying. In the Non Objective 1 region 28% are in dairying. The shifting relativities between dairying and beef production – especially the suckler cow enterprise – can be seen from the changing ratios of other cattle to dairy cows in each of the regions which show a regional pattern.

The increase in the ratio of other cattle to dairy cows was somewhat greater in the Non Objective 1 areas than in the other regions; specialist beef farms increased at a faster rate in the former. Nevertheless, the southwest remains a comparatively strong dairying region and dairy farms generally are still the largest production units in economic terms. Nationally, in 2000, specialist dairy farms had an average of 45.8 ESUs – more than double the national figure. Also, dairy farms had an average of 1.07 ESUs per hectare, compared to 0.39 ESUs per hectares on beef farms. Clearly enterprise shifts have favoured the southern regions – where dairying decline has been slower and increases in other cattle relatively greater.

Following the 1992 CAP reforms, an *early retirement scheme* for farmers was introduced in Ireland in January 1994. Under the scheme, farmers aged between 55 and 66 years are eligible to participate if they have practised farming as their main occupation for the preceding ten years. Unlike previous retirement schemes the level of guaranteed pension payments were sufficient to stimulate a high level of participation. Over the first five years almost 8% of all farmers had participated (Gillmor, 1999). Most significantly, there was a very distinctive spatial pattern to the adoption of the scheme, which has persisted to the present, with the highest levels in the more prosperous farming regions where there is a substantial number of trained young farmers who qualify as suitable transferees. With a less favourable resource base, farm structures and demographic situation in many parts of the west and northwest, the level of adoption has been much lower (Laffery *et al.*, 1999).

The influence of direct payments on farm incomes in Ireland increased significantly in the aftermath of the EU CAP reforms of 1992. Many aspects of the timing of the payments are left to the discretion of the Member States, so that they may be delayed or brought forward, with a consequent bearing on farm incomes in any particular year. In 1992, the last year before CAP reforms, direct payments accounted for just over 30% of average family farm income. This had increased to 60% by 1996 and to 90% by 2002 (Frawley and Phelan, 2002). Their impact was strengthened by the provisions of Agenda 2000, but their proportionate contribution to family farm income is also influenced by changing levels of producer prices.

Direct payments can account for more than 100% of farm income when market-based output is not sufficient to cover total costs. In 2000, on cattle and sheep farms, direct payments represented about 120% of family farm income. The concept of direct payments as a proportion of income has not had the same relevance for dairying as for the other major systems as these payments are not used as a mechanism under CAP for supporting dairy farm incomes. Consequently, such payments account for only about 20% of farm income on specialist dairy farms (Connolly *et al.*, 2001, p. 8).

Regionally, the impact of DPs on family farm income is highest in the Border counties, the Midlands and the West, where they constitute the greater part of farm income (Table 5.4). These are the regions in which cattle rearing and sheep are predominant enterprises.

Table 5.4: Direct payments as a percentage of family farm income, Ireland, 1996–2002

Region	1996	1997	1998	1999	2000	2001	2002
Border	78	78	97	106	89	93	119
Midlands	73	82	90	85	91	89	101
West	78	74	86	108	91	88	118

Table 5.4 (cont)

Region	1996	1997	1998	1999	2000	2001	2002
East	61	61	71	76	67	63	95
Mid-West	43	48	62	67	72	64	72
Southeast	55	57	58	61	57	63	79
Southwest	39	47	52	53	45	52	71
State	59	62	69	74	68	72	90

Source: Teagasc National Farm Survey, various years.

5.8.3 Non-farm employment and income sources

The Irish Census of Agriculture asks farmholders to state the degree of importance that farm work as an occupation holds for them. Three categories are specified: sole occupation, major occupation, or subsidiary occupation. In 2000, the distribution nationally was as follows (with corresponding figures for 1991 in brackets): sole – 55.7% (73.4%); major – 13.9% (5.7%); and subsidiary – 30.4% (20.8%). While there was little variation among the NUTS2 regions, county-level data show higher percentages of sole occupation farmholders in the larger farm areas of the south and east.

Between 1991 and 2000 the major shift was from "sole" to "major" occupation status (Table 5.5). The change to "major" status was least pronounced in the west but in 1991 the West region had already a relatively high percentage of farmholders in the category and maintained this distinction in 2000. The changes recorded are influenced not only by farming conditions but by the distribution of non-farm employment.

Table 5.5: Importance of farm work: changes (%), Ireland, 1991–2000

Region	Sole Occupation	Major Occupation	Subsidiary Occupation
Border East	−37.7	+143.1	+13.2
Border West	−37.3	+75.1	+19.1
Midland	−36.8	+188.0	+24.0
West	−37.8	+55.2	+28.8
Dublin	−45.3	+37.8	−41.1
Mid-East	−33.4	+85.4	+3.8
Mid-West	−37.2	+127.4	+28.1
Southeast	−34.6	+200.4	+20.0
Southwest	−37.6	+106.3	+26.3
Objective 1 region	−37.5	+85.4	+23.0
Non Obj. 1 region	−36.3	+124.4	+19.3
State	−36.9	+101.6	+21.4

Source: Census of Agriculture 1991 and 2000.

The data in Table 5.5 only provide an indication of the diminishing importance of farm work; they do not give a direct account of the movement from farming to other occupations. Data from the Teagasc (The Agricultural Research and Advisory Authority) annual National Farm Survey (NFS) show the trend to off-farm employment among farmholders and their spouses/partners. Results for 2002 indicate that, on 48% of farms, the farmer and/or spouse had another occupation. For farmholders separately the figure was 35%. Since the early 1990s the trend towards off-farm employment has been gradually upwards. Its extent varies significantly by system of farming, with dairy farmers being much less likely (12%) and cattle rearing farmers more likely (49%) to combine farm and non-farm work.

Farmers with other occupations tend to have smaller farms – 27 hectares on average – with fewer livestock and lower stocking density that those without off-farm jobs. Direct (non-market) payments make up a higher percentage of their family farm income, and they have lower incomes from farming. The National Farm Survey data show regional variations in off-farm employment (Table 5.6). The incidence of part-time farming among farmholders is highest in the West, Mid-west, Midlands and the Border region (which extends to the north-west).

Table 5.6: Off-farm sources of income: farmholder and/or spouse, Ireland, 2000

Region	Holder and/or spouse (%)	Holder (%)	Spouse (%)
Border	41.9	32.3	21.4
Midlands	54.9	39.1	25.7
West	56.4	48.6	28.1
East	38.4	22.5	22.0
Mid-west	54.0	37.0	30.2
Southeast	34.9	18.6	21.3
Southwest	38.2	19.5	24.6
Total	45.0	32.7	23.0

Source: Teagasc National Farm Survey 2000, Tables 7E and 14E.

The national Household Budget Survey (HBS), conducted on a representative sample of private households every five to seven years, provides information on all sources of income to a household – as well as on household expenditures. The latest HBS relates to the period June 1999 to July 2000. In the HBS "farm households" are defined as those where the head of the household is:

- a gainfully occupied farmer, or
- a retired farmer with at least one gainfully occupied farmer in the household.

This definition does not include other households involved in farming (e.g. where farming is a subsidiary occupation of a head of household who has a main occupation outside farming). Table 5.7 shows the contributions of the major sources of "Direct Income" to rural farm households. These data are not available for regions.

Table 5.7: Components of direct income in farm households, Ireland

Source	1987 (%)	1994 (%)	2000 (%)
Wages/Salaries	29	35	48
Farming	59	58	44
Other	12	7	8
Total Direct Income*	100	100	100

*Employment and other income but excluding state transfers
Source: Household Budget Survey 1999–2000.

Wages and salaries accounted for 48% of Direct Income in farm households in 2000, compared to 29% in 1987. The increase in this proportion between 1994 and 2000 is particularly remarkable, and coincides with the so-called "Celtic Tiger" period of growth in the Irish economy (Commins, 2003).

When considering income relativities across different categories of household, it is more appropriate to refer to Disposable Household Income (DHI) obtained by adding state income transfers to Direct Income (to give gross household income), and subtracting direct taxation payments. On this basis, Table 5.8 compares the 2000 position in farm households with that in "rural non-farm" and "urban" households

Table 5.8: Household income relativities, Ireland, 2000 (State=100)

Category of Household	Direct Income	State Transfers	Gross Household Income	Direct Taxation	Disposable Household Income
Farm	97	83	95	57	103
Rural Non-Farm	79	109	82	69	85
Urban	110	98	108	119	106
State	100	100	100	100	100

Source: Derived from Household Budget Survey 2000.

For farm households, both Direct Income and Gross Household Income were close to the State average and higher than in the rural non-farm category. Furthermore, the operation of the national taxation system favoured farm households (because of lower personal incomes, see below) with the result that Disposable Household Income on farms exceeded the State average and came close to the Urban average.

This analysis, however, ignores household differences in the number of persons who are economically active and thus contributing to household income. Farm households have more persons, on average, at work than other categories of households. When account is taken of this, by dividing direct incomes from employment (including income from farming) by number of persons at work, it turns out that the relativities between categories of household change substantially. The ratio (to the State average 100) in respect of disposable income for farm households drops to 73 while those for rural non-farm and urban households rise to 90 and 109 respectively (Commins, 2003).

It would appear that over time, and with general economic growth, rural farm households have acquired more earners than other types of households. However, taking into account their low per person earnings and their low absolute levels of direct taxation, it seems that their work – whether in farming or outside it – yields relatively low incomes.

Additional analyses have been undertaken comparing farm incomes with the levels of earnings outside the farm (Commins, 2003). These show that family farm income per family labour unit on full-time farms has, since 1987, at best, remained constantly at 20% lower than average male earnings in manufacturing industries. Even in "good farming years", farm incomes did not reach 90% of industrial income. In this context and with employment levels generally approaching "full employment" status, it is clear that workers in farming households, except on the more commercial farms, will be "pulled" towards the non-farm labour market.

5.8.4   Rural enterprise and employment

This section refers to "the broader rural economy" and reviews trends at macro level with special reference to the record of State-sponsored agencies in creating employment so as to offset labour losses in farming. With the unprecedented growth in the economy generally over the past decade, the concern for "balanced regional development" came strongly into focus. This, coupled with the launch of a National Spatial Strategy, indicated that with the national economy reaching near-full employment there has been a new impetus towards territorial development in the rhetoric of official policy. The question that arises is: what has been the experience of the more rural regions in regard to enterprise and employment during the period of national economic growth in the 1990s?

The Gross Value Added (GVA) index points up the poorer regions of the Border counties, the Midlands and the West, which together constitute Ireland's Objective 1 region. When employment income per person at work is considered, these regions improve their relative position but, surprisingly, the standing of the Southeast worsens (as does Dublin's). Data on disposable income per capita show the Midlands and Southeast to have the lowest rates of increase between 1995 and 2000 and, on the criterion as an index for 2001, to be the two most disadvantaged regions in the country. The similarity of these two regions with the Border and West regions lies in their comparatively high dependence on primary production.

In the "boom years" up to mid-2001, two regions combined, Dublin and the Mid-East, gained 45% of the extra employment created. All regions shared in the gains but the rates of employment growth were weakest in two of the country's most rural regions – the Border and Midlands. In the third predominantly rural region, the West, the rate of employment growth (+48.7%) greatly exceeded the national average increase of one-third but was most heavily concentrated in the sub-region centred on Galway city.

Until the small-area returns from the 2002 Census are examined in detail, there can be no definitive assessment of the intra-regional disposal of recent employment growth. However, judged on the basis of trends between 1991 and 1996 – before the significant expansion of the late 1990s – the spatial distribution of growth is likely to have been quite widespread (Commins and McDonagh, 2002). A summary of the basic facts for 1991–96 follows:

- The combined 155 Rural Districts in the State had a net loss of 22,400 persons at work in the primary sector, but this was offset by a gain of 100,230 workers in other sectors.
- Only a minority of Rural Districts, mainly in the West and in dispersed upland areas, did not have sufficient non-farm employment growth to counter losses in the farm workforce.
- Most (88%) of the State's 3,421 District Electoral Divisions had an increase in the number of self-employed outside the primary sector.
- When analysed by size of place the number at work increased for all categories of place, although places below 1,500 persons had the lowest rates of increase.

Among the State's programmes for supporting enterprise and job creation, foreign-owned enterprises have been the main generators of employment, and can account for more than half of the jobs established in enterprises supported by State Agencies. Relevant data are as follows:

- The longer-term trend (1981–98) in the establishment of State-funded enterprises confirms a significant increase but the East region (Dublin and hinterland counties) gained very substantially, doubling its share of new foreign firms between 1981–86 and 1993–98.
- In the case of new indigenous firms, the regional distribution has remained more balanced, although the East again has increased its share.
- The East also gained disproportionately in the number of gross job gains. The East accounted for just under half of the expanding numbers of jobs in 1993–98, compared to one-third in the 1980s. The East also increased its share of gross job creation in indigenous firms.
- With regard to the locational distribution of employment and enterprise within regions, there is a definite trend towards centres of over 5,000 persons.
- The outcome of these centripetal tendencies in grant-aided enterprises is that by the end of the 1990s about one-sixth of firms assisted by the main State agencies were located in the more rural areas (places below 1,500 persons). By contrast, over half were in the centres of 10,000 persons upwards. Employment creation patterns mirror those for enterprise development.
- Related to these trends is the fact that the fastest growth sector has been the traded and financial services. These expanded employment by 50,000 in 1991–2000, with over two-thirds of this employment being provided by foreign firms, predominantly located in the city regions.

The evidence presented above shows that, while all regions benefited from the economic boom of the 1990s, there has been a clear tendency for enterprise and jobs supported by mainline agencies to become located in the eastern part of the country and, within regions, in the larger centres of population. Two programmes are expressly concerned with the promotion of SMEs at local level. These are the County Enterprise Boards (CEBs) programme and the LEADER programme.

The CEBs commenced activities in 1993/4, having been set up to cater specifically for micro-enterprises. By 2000, 34 Boards had approved a total of 14,000 projects, one-third of which were in the Objective 1 Region. CEB-assisted new enterprises and expansions resulted in the creation of 21,500 full-time jobs and 4,800 part-time jobs. When counties were classified by degree of "rurality", analysis of CEB records showed job provision with a clear bias in favour of more rural counties (Table 5.9). Highly rural counties, with 16% of the State's population, accounted for 25% of grants drawn down and jobs provided.

Table 5.9: Grants drawn and jobs created in CEB enterprises, 1993–2000, in counties classified by degree of rurality, Ireland

	High*	Medium*	Low*	All
Grants Drawn %	25.1	30.6	44.3	100
Jobs Created %	24.5	32.9	42.6	100
Population %	16.0	26.9	57.1	100
Jobs per 1,000 pop.	9.1	7.3	4.4	100

*Based on population distribution by size of place. "High rurality" is equated with at least 75% of the population residing outside centres of 3000+ inhabitants. "Low rurality" corresponds with less than 50% residing outside such centres.
Source: Commins and McDonagh 2002.

The LEADER programme has been an important catalyst for rural development because of its accessibility at local level, and its accountability to local people through its partnership structures. The second phase of the programme ran from 1995–99, and involved 34 local groups. This phase had covered 9,595 projects, creating 8,357 full-time job equivalents distributed over several categories of activities. Three categories are of relevance here: (i) rural tourism; (ii) small firms, craft enterprises and local services; and (iii) agriculture, forestry and fishing, which together account for 4,324 of the total projects undertaken by the local groups: 55% were in rural tourism, 30% in small businesses/services, and 15% in agriculture forestry and fishing.

As in the analysis of CEB activities, LEADER groups were classified into three categories, again on the basis of the degree of rurality of their catchment areas. There is a clear gradation from "high rural" to "low rural" in the amount of funds allocated per 1,000 population to LEADER groups. Correspondingly, there were clear differences between the different categories of rural area as regards expenditures and projects undertaken per 1,000 population, except in activities under "agriculture, forestry and fishing". Seemingly, in the more rural areas, tourism and small businesses development offer most potential to LEADER groups.

Data on LEADER groups within their catchment areas are limited. On the basis of an examination of some case studies, it appears that there has been a tendency for concentration around county towns (Kearney Associates *et al.*, 2000, p.79). This is perhaps understandable given the concentrations of population and, possibly, greater economic opportunity in those areas. An additional feature of the LEADER programme is that it has encouraged innovation and enhancement of social capital in many areas that were previously neglected by public agencies (Walsh, 1997, 1999b).

5.8.5   Concluding remarks

The basic lesson from the Irish case study may be summarised in the following terms. The territorial impacts of agricultural and rural development policies vary with the aims of such policies but are also differentiated according to the resource and structural characteristics of regional economies. Secondly, there is a longer-term techno-economic process of agricultural restructuring onto which policies are layered. Policies may cushion the more deleterious impacts of this on farm households (e.g. by supporting incomes) and thus slow the rate of structural change, or "go with the flow" and facilitate desirable adjustments (e.g. by promoting alternative forms of land use). Thirdly, policies, when considered in their totality, may have inconsistent outcomes – as for example when farm price policies and even direct payments have territorial impacts that run counter to cohesion objectives.

Finally, it is clear from the Irish case study that in the more commercially oriented farming regions a comprehensive range of agricultural policies, and/or farm-centred rural development policies, do not provide a guarantee of rural demographic viability. There is a need for greater complementarity between agricultural policy measures and policies for broader regional development focused on the specific conditions of the different regions.

5.9. Country case study: Poland[18]

Whereas the Polish pre-accession SAPARD programme built on experience from similar national and externally funded support, the degree of change in both the manner and form of support required new administrative systems to be put in place, as well as the adoption of EU principles for development. As such, the experiences of the SAPARD programme in Poland are realistic and valuable evidence of the essential ingredients of successful rural development actions both in the design of activities and in implementation.

---

[18] This section is largely based on Dalton (2004).

Above all else, the SAPARD has been a steep learning experience for all concerned. For example, although subsidised farm credit has a long history with well-known procedures and institutions, capital grants were a new way of supporting the sector in Poland. Extensive delays occurred in the delivery of the Polish Programme, due to the time taken to learn about and put in place new management systems, and also due to delays in the uptake of assistance once it became available. These delays reinforced the emphasis on the political objective of spending the funds prior to accession at the expense of achieving the intended impacts of the Programme.

There were two balanced priority axes within the Polish SAPARD Programme. Priority 1 was to improve the efficiency of the agro-food sector, and comprised Measure 1 (improvement in the processing and marketing of food and fishery products) and Measure 2 (investments in agricultural holdings). Priority 2 was to improve business conditions and job creation, and comprised Measures 3 (development of rural infrastructure), 4 (diversification of economic activities in rural areas), 5 (a pilot agro-environmental scheme), 6 (vocational farmer training schemes) and 7 (technical support). The original financial allocation by measure was a negotiated equilibrium between the different interested parties at the time including those of the existing Member States. The allocation of funding by measure also reflected different judgments concerning priority weights, expected rates and speed of uptake. For example, Measure 3, improving rural infrastructure, was rightly predicted to be the measure most easily taken up, and thus was given a higher proportion of funds in the first years of the 7 year Programme.

Agrarian interests within Poland can be characterized as those representing the "farm" as opposed to those representing the "village". Due to the new way that consultation about the design of the programme took place through regional seminars using the new regional government structure, no particular lobby group could claim to have been overlooked. Many of the subsequent changes to the Programme implemented by the Monitoring Committee sought to balance some of these interests as it became clear that expenditure on rural infrastructure was the most readily taken up. For example, after some modification, Sub-Measure 2.3 (incentives to adopt diversification and first-stage on-farm processing) became the most popular farm sub-measure.

The deadline of submitting the Programme by the end of 1999 was met and was subsequently approved by the Commission in September 2000, with the expectation that the Programme would be enacted quickly. These hopes were dashed when the extent and uncertainties of the required effort to design and agree all the procedures and accredit the devolved managerial and operational authorities gradually became apparent. This process took until July 2002 when the multi-annual financial agreement was finally signed along with the annual financial agreements for 2000 and 2001. The actual result of this process was a rigorous set of most detailed procedures and institutional responsibilities that were formally set out in great detail. The outcome was that the Programme was finally begun on 9 July 2002, but only in part. Notably, Measures 4 and 5 had not begun by November 2003 even though exhaustive accreditation procedures had been completed for Measure 4.

One commonly held perception was that the application and evaluation procedures and the eligibility criteria of the Programme were complicated and difficult, a fact supported by the slow rate of applications right up until the autumn of 2003. The main concern of the Programme managers up until this point was whether or not the allocated funds could be taken up (absorbed). Decentralised management was also a new experience for Commission officials, especially when the financial procedures of the FEOGA Guarantee Section were adopted despite much lobbying by the Polish authorities. It meant that no advance payments were available and, initially at least, no multi-annual funding commitments. The accreditation of the management authorities was a long and arduous process. While the needs analysis and the objectives of the Programme were used to design the measures, eligibility and evaluation criteria were primarily driven by the operational requirements. Moreover, the process of changing these procedures is perceived as being so energy- and time-consuming that the SAPARD system could be said to be both unworkable and at the same time unchangeable. Despite this, a number of resolutions were passed and implemented by the Monitoring Committee, and, in addition, the procedures have been an effective screen against poorly prepared applications.

In addition to the effort that went into the consultation exercise to provide the basis for the Programme, conforming to the principles of the EU structural funds – programming, co-financing, partnership and targeting as well as monitoring and evaluation – resulted in significant innovations. The achievement of national co-financing at a time of stringent measures to control budget deficits is noteworthy. Co-financing by the beneficiaries makes the beneficiaries more responsible for their actions but has been used as a criticism of the Programme and is often assumed to be a reason for slow uptake. The financing problem was moderated by the greater possibility of paying claimants in instalments rather than simply when the investments have been completed, and also by increasing the rate of grant up to the maximum 50% rate for most measures and sub-measures. Total eligibility expenditure maxima were also raised substantially, especially in the case of Measure 1 (food processing).

The application of the partnership principle was boosted by the installation of a regional system of government in Poland in 1999. Each of the 16 Polish regions was consulted in the Programme formulation process. While more direct involvement in Programme management by regional authorities was not implemented, the regional steering committees have a role in the determination of the selection criteria for Measure 3 and in the ranking of these rural infrastructure projects (which became meaningful in that the number of applications for this measure exceeded the funds available). In addition, in view of the divergences between regions, a national steering committee advised on regional funding envelopes so as to ensure that resources were not allocated unfairly.

The Polish SAPARD agency (the Agency for Reconstruction and Modernisation of Agriculture, ARMA) established regional offices with a total staff approaching 400 devoted to the SAPARD Program plus a further 200 staff in the head office. There was a single paying agency, which in view of the costs of accreditation is a sensible and cost-effective situation. Some functions such as the evaluation of Measure 2 projects were outsourced to the National Agricultural Advisory Centres (NACARD) – the national extension service. Targeting was used extensively in the design of the measures, although not in any significant geographic way.

In the Polish agricultural budget for 2002, the value of these subsidies amounted to some 236 million Polish zloty or roughly 59 million euros. These subsidies were targeted at specific structural issues including many measures similar to those included in SAPARD, including infrastructure, the modernisation of processing plants and rural job creation, but also support for organic farmers and for young farmers to establish farms. The World Bank rural development programme includes a micro credit scheme although one component also includes loans for infrastructure in objectively identified regions. Thus, at the start of the Polish SAPARD, there was a body of expertise concerning public assistance in the areas where SAPARD was focused, albeit on a different basis and with different procedures.

The analysis of needs in the Operational Programme pointed up the duality of the Polish agricultural situation, i.e. the poor agricultural holding structure and low incomes from agricultural activities. Financing problems faced by farmers also featured strongly in the arguments for assistance. The need to adopt the *acquis* in the dairy sector, given the drive for quality milk, was given special emphasis and lay behind the special dairy sub-measure (2.1) Animal welfare enhancement needs were the main basis of other agricultural sub-measures. Given the constraint of oversupplied agricultural markets both in Poland and the EU, support was also justified for the production of non-traditional enterprises and for adding value on the farm through first-stage processing. This sub-measure (2.3) proved to be by far the most popular in Measure 2.

The most important reason for supporting the food-processing sector was the need to have in place by accession plants licensed to have reached the standards of the *acquis* and so able to trade their products in the whole of the EU. This was a sizeable and expensive task in a sector with low profitability and much concentration and structural change taking place against a background of dynamic changes in the nature of the demand for more processed and higher quality food.

The poor state of rural infrastructure was a central argument for Measure 3, as shown by a variety of indicators of the availability of roads, water supplies, drains and waste disposal facilities. The poor situation for social capital, especially for farmer education and training underpinned Measure 6, while the severity of the unemployment and under-employment situation gave strong evidence for employment creation measures to be supported in Measure 4. This was one of the more important rural-urban disparities addressed within the SAPARD Programme.

The agro-environmental measure was for a pilot scheme along with a forestry sub-measure and was justified on the grounds that some experience of these accompanying measures prior to accession would be most valuable after accession when they would need to be made available on a national basis.

Targeting of beneficiaries was a major focus in the Programme, as the expectation was that insufficient funds were available for the potentially very large number of beneficiaries. Thus the focus on the *acquis* defined to a large extent the need to support specialist farms in their adoption of more appropriate buildings and facilities for keeping animals and for the storage and disposal of manure. Viability criteria featured strongly in conditions for eligibility including the intention to develop, an upper age limit of 50 years, which was subsequently raised to 55 and proven experience of farming and/or qualifications to do so. Viability checks were also important in the selection of eligible food-processing entities including size constraints and information on the security of their sales, their financial situation and the quality of their management.

Rural *Gminas* (local communities) and associations of *Gminas* were the targeted beneficiaries for Measure 3 on support for infrastructure. Local municipalities could also be beneficiaries for the restoration and enhancement of local tourist facilities in Measure 4. Farmers and their family members plus rural entrepreneurs were the targeted beneficiaries for employment creation initiatives.

Despite the big differences in farm and food-processing business size and their very different financial situations, single measures for all types of business were put in place. Similarly, despite the evidence of the strong urban and rural distinction in welfare no direct provision was made in Measure 3 for this situation. However, a system of financial allocations for each region (regional envelopes) was designed in order to prevent funds being allocated in an "unfair" way to any single region and criteria were agreed on the basis for this allocation.

From an administrative viewpoint, the appointment, establishment, accreditation and activation of the necessary institutions proved to be greater tasks than ever anticipated. Unfortunately this process did not begin until after the Programme was agreed which meant that obvious institutional constraints were not taken into account in the Programme design. For example, Measure 4, which had large numbers of small potential applicants, would have been difficult to service even if it had been launched at the same time as the other measures simply because of a lack of staff and the exhaustive and detailed nature of the application and approval procedures.

The evaluation also pointed up a lack of balance between the various issues in programme management concerning possible malpractice and administrative costs. All projects irrespective of their size were to be visited twice which took up about 30% of total processing time. Evaluation of Measure 2 was simply a paper exercise where the evaluator was not allowed to visit the farm. There was no provision for a simplified application process for small amounts of assistance in any of the measures.

Another example of "micro management" was that in the procedure to change procedures all suggestions were referred to the EU Commission irrespective of their importance and the principle of delegated management. This resulted in both delays and the situation that even the most sensible proposed change was perceived by staff as too difficult to achieve. The main losers from this failing were ultimately the beneficiaries as the whole Programme was delayed. Indeed a significant omission from the multi-annual financial agreement was a procedure to ensure that such delays in the whole Programme did not occur. No institution seemed to have the responsibility of making sure the Programme kept to a time schedule even though it is easy to show that such delays are very costly.

In conclusion, the main achievement of SAPARD in Poland was that it was successfully implemented despite a huge and steep learning curve for all concerned. Given accession in May 2004, 4 years of funding were almost committed within a period of 20 months. One of the explanations of how this was achieved was the fact that there was considerable experience to build on from previous initiatives of this type and a few key staff involved in these were recruited to lead the SAPARD effort. However, to simply adopt this measure of success ignores the fact that some important parts of the Programme were not implemented, which inevitably warped the coherence of the plan. Moreover, the main observed goal of the Programme was to spend the available funds. Achieving the goals of the Programme was not so important. The information required of the monitoring system and the information used and decisions made by the Monitoring Committee reflected this emphasis. In an effort to spend the funds, the Programme was steered more towards meeting the private interests of beneficiaries. This fact is very strong evidence for planning the contents of a Programme according to both needs and implementing capacity. The Polish Programme also demonstrated the importance of considering the implications for processing capacity when changes are made.

The main findings of the case study are that the uptake of farm and rural development support in New Member States depends heavily on a suitably designed Programme from several different perspectives, namely the needs that are to be addressed, the ability of the intended beneficiaries to access the assistance, and the capacity of the administrative system to process applications on time.

Chapter 6

6. The CAP/RDP Reforms in the EU-15 and the NMSs

6.1. Introduction

This chapter seeks to build on the attempts of others to model the impact of recently agreed CAP reforms. Any kind of territorial impact assessment of a policy faces the basic methodological problem of separating the effects of the policy from the effects of other factors influencing complex spatial structures. Thus more sophisticated methods than were used in Chapter 4 are necessary to assess "impacts" as opposed to simply measuring the "incidence" of CAP support as was done in the earlier chapter.

In July 2002, as scheduled in the Agenda 2000 decisions, the Commission brought forward its Mid-Term Review of the CAP (CEC, 2002). The Review included a number of earlier proposals, which were re-stated in more detail and with some modification in the later Explanatory Memorandum to the Commission's Long-Term Policy Perspective (CEC, 2003) for agriculture. In June 2003, the Council of Agricultural Ministers reached agreement on a further major reform of the CAP.

This chapter summarises existing analyses of these reforms (mainly the original MTR proposals of July 2002, since little subsequent analysis has yet been widely reported). It starts with a description of the Commission's MTR proposals (which do not vary widely from the later package agreed by the Council), and reviews some basic analysis. The following section shows the results of quantitatively analysing at a NUTS3 level some estimates from one of these studies (CAPRI). Finally, the implications of CAP reform for the New Member States in Central Europe are discussed.

6.2. The MTR proposals, agreement and basic analysis

The MTR proposals involved the following main points:

- *Crops*: In the July proposals, compulsory long-term (10 years) set-aside on arable land (replacing rotational set-aside) to form part of "cross-compliance" (see below). Support for non-food crops to take the form of a carbon credit, a non-crop-specific aid worth €45 per hectare of energy crops up to a maximum of 1.5 Mha. The January 2003 proposals added a 5% cut in the intervention prices of cereals, with an increase in direct payments for cereals and oilseed areas, and a new payment system for protein crops.

- *Livestock*: Milk quotas maintained until 2014/15. Agenda 2000 intervention price cuts to be introduced one year earlier (i.e. in 2004) and extended to 2008, with asymmetric cuts in skim milk powder (–3.5%) and butter (–7%) and an increase in quotas. No specific proposals for beef, etc.
- *Single Farm Payment*: To replace all existing direct payments to producers, with a number of exceptions (e.g. durum wheat, rice), and be based on historical levels of payment to each farm. Payment to be subject to a number of statutory environmental, food safety and animal health and welfare standards, as well as occupational safety requirements for farmers. This "cross-compliance" should reflect regional differences, and distortion of competition was to be avoided by means of a "common framework providing basic implementation criteria" within which Member States would define and enforce standards on a whole-farm basis. A compulsory farm audit for all commercial farms receiving over €5000 per year in direct payments.
- *"Dynamic [or degressive] modulation"*, to reduce all direct payments by 3% per year to reach 20% (the maximum agreed in Agenda 2000). However, a franchise of €5000 of direct payments to be applied to all farms with up to 2 full-time annual work units (AWUs) plus €3000 for each additional AWU. This would exempt around three-quarters of all EU-15 farms but affect under a fifth of all direct payments. A "capping" maximum of €300,000 in direct payments to apply to all farms. In its January 2003 proposals, payment totals over €5000 but below €50,000 were to be cut by steps from 1% in 2006 to 12.5% in 2012, and by steps from 1% to 19% for payment totals over €50,000.

Rather than (as previously) allowing Member States to spend funds made available by modulation within their own accounts, funds saved by the June 2002 proposals were to be distributed from the EU budget "to Member States on the basis of agricultural area, agricultural employment and a prosperity criterion, to target specific rural needs". This was expected to "allow some redistribution from intensive cereal and livestock producing countries to poorer and more extensive/mountainous countries, bringing positive environmental and cohesion effects" (CEC, 2002, p. 23). However, savings from capping would be redistributed according to the amount capped in each country. All such funds saved from Pillar 1 would be used by Member States to reinforce Pillar 2 rural development programmes financed under the FEOGA Guarantee section. In the January 2003 proposals, the first 6% of these savings would be transferred to Pillar 2; the remainder would be used to finance future market needs.

Rural Development Policy would "consolidate and strengthen the Second Pillar by increasing the scope of the accompanying measures and widening and clarifying the scope and level of certain measures" (CEC, 2002, p. 24). New measures were to include new chapters on food quality and on meeting farming standards, and introduce animal welfare payments into the agri-environment chapter.

The CAP reforms actually agreed on 26 June 2003 involved the following:

Crops: No change in the cereal intervention price, but a halving of monthly increments, i.e. a small effective reduction (but no additional compensatory payments). Minor changes in the regimes for rye, protein crops, rice, durum wheat, nuts, starch potatoes and dried fodder. A "carbon credit" energy crop aid of €45/ha, to a maximum of 1.5 Mha.

Milk: The intervention price for butter to be reduced by 25% over 4 years, i.e. an extra 10% cut compared to Agenda 2000 cut; and that for skimmed milk powder by 15% over 3 years as previously agreed, i.e. an overall cut of about 20% for milk. Only minor quota changes other than the Agenda 2000 increases scheduled for 2006 onwards. Direct dairy payments (agreed in Agenda 2000) to be introduced as scheduled (i.e. from 2004), but kept separate from the single payment (see below) until 2008 at the earliest. This implies that the dairy regime is to be maintained in roughly its present form for the foreseeable future.

Single Farm Payment (SFP): From 2005 (or 2006 or 2007), a payment direct to EU farmers, to be based on historical (2000–2002) receipts (less 3%) of arable and livestock payments, but independent of (i.e. "decoupled" from) levels of farm output or resources (land area, livestock numbers, etc.). Eligible land (i.e. land with SFP "entitlement") is all arable land and grassland, except land on which fruit, vegetables or table potatoes are grown, and land in permanent cropping (short-rotation coppice etc. is not regarded as permanent). This land need not be that from which the entitlement was first established. Member States may redistribute SFPs within regions, e.g. via uniform (flat-rate) payments per hectare, or separate aid rates in each region for permanent pasture and cropland. Existing set-aside obligations will continue. Entitlements may be transferred (e.g. sold or leased, after some use) to those with sufficient agricultural land, within national and possibly regional boundaries.

Recoupled Payments: However, in order to avoid destabilising the present farming structure too much, Member States may retain up to 25% of arable payments, up to 50% of sheep and goat premiums (including LFA supplementary premiums), and up to 100% of suckler cow premium (on various bases). It appears possible that these retained payments may be re-allocated on a somewhat different – e.g. regionally differentiated – basis from that used to date.

Member States may also make additional payments, at national or regional level (but without co-funding), to encourage specific types of farming which protect/enhance the environment or to improve quality and marketing, up to 10% of national sectoral expenditure ceilings (arable, beef, sheep, dairy). The SFPs will then be reduced correspondingly.

The SFP will be linked to the maintenance of standards of environmental care, food safety, animal and plant health, and animal welfare, and the requirement to keep all farmland in good agricultural and environmental condition ("cross-compliance"). Farm advisory services will become compulsory in Member States by 2007, although farmer participation will be voluntary.

SFPs above a "franchise" level of €5000 will be reduced ("modulated") at a single flat rate of 3% in 2005, 4% in 2006 and 5% from 2007 onwards, in order to finance rural development policy by about €1.2 billion by 2007 onwards. At least 1% will be re-distributed to the Member State, and the rest according to a Commission key, but Member States will receive at least 80% (Germany 90%, the extra for rye regions) of their "own" modulation funds. New Member States are exempted from modulation and its financial (budgetary) effects until their levels of direct aid align with EU-15.

Rural Development Policy will be strengthened with more EU (modulation) funding from 2005 onwards for new measures and/or extra funding for: the environment (with higher Community contributions), food quality, young farmers, animal welfare, and to help farmers meet EU production standards.

Financing: If the CAP budget (subheading 1a, i.e. Guarantee) fixed to 2013 is considered by the Commission likely to be overspent, then direct payments will be reduced, but not to farmers below €5000 (and perhaps smaller reductions for additional, higher, franchises).

A number of "impact analyses" of recent CAP reform proposals have been undertaken, several by or at the initiative of the Commission. Four were reported in *Rural Development in the European Union* (EC, 2003b), along with the Commission's own studies, all comparing simulation estimates of (i) the situation in 2009 if the MTR proposals were implemented (with no other changes in the CAP or in macroeconomic conditions, but usually with exogenous assumptions about labour, land and other productivity trends from a recent base period) with (ii) the estimated "reference" situation for the same year if the CAP were unchanged (except for complete working out of the agreed Agenda 2000 reforms). The results, in percentage changes for physical amounts and prices, and sometimes in Euro for income and welfare effects, are thus "comparative static" in nature, isolating the MTR impact from other influences on agricultural performances. None of the models simulate the non-agricultural rural sector explicitly, and several are purely agricultural in nature.

The Commission itself carried out two studies of the MTR proposals, one using its standard set of partial-equilibrium (i.e. agriculture only) dynamic models used for regular market outlook work, and the other using the ESIM agricultural sector model originally developed to study the implications of EU enlargement. Compared to the reference situation, MTR implementation is estimated in the first study to reduce the area of cereals, oilseeds and fodder crops but to increase that of set-aside energy crops. Beef cattle numbers and output would decline, but prices would rise. Overall, factor income (GVA plus subsidies) would be almost unaffected, if it is assumed that most of the modulation savings are returned to farmers via Pillar 2.

6.3. Territorial analysis of CAPRI study impact results

The CAPRI modelling system was developed at the EuroCARE centre at the University of Bonn and elsewhere as a FAIR3 project in 1997–1999, and has been updated and improved in a more recent project, CAP-STRAT. The main objective of these projects was the development of an EU-wide economic modelling system able to analyse regional impacts of the CAP. In the following, a short description of the CAPRI modelling system is given before the results of the MTR study are presented (see Britz, 2004 for a more detailed description).

The modelling system consists of a regionalised database and a corresponding core model that is strictly in line with micro-economic theory (van Tongeren, 2004). The core model consists of a supply module (for EU-15) of 200 sub-national regions at NUTS2 level and a market module of the EU-15 and 11 non-EU global regions (e.g. the U.S., Canada, Australia/New Zealand, India, China, NMSs). The system features a detailed regional description of CAP measures including payment schemes, set-aside obligations and quotas on the supply side, and price floors, market interventions, tariffs including tariff rate quotas and bilateral trade agreements as well as export subsidies on the market side. For non-EU regions, policies are based on OECD's PSE/CSE database.

The CAPRI system involves physical consistency balances for agricultural areas, young animals and feed requirements for animals as well as nutrient requirements for crops. The production activities are detailed, and in accordance with the economic accounts for agriculture (over 60 products and 30 inputs). Each region can be seen as a "farm" that maximises its profit function by choosing the optimal composition of inputs and outputs, at given prices for the final product and given prices for key inputs (van Tongeren, 2004). The sub-national "farms" (at NUTS2 level) are then aggregated to Member-State levels using techniques from maximum-entropy and positive mathematical programming. Trade occurs between Member States, and market clearing at the EU-15 level yields prices for inputs and output (including feed and young animals). Through an iterative procedure, supply is optimised for each NUTS2 region, and new market prices are calculated until the whole system is in equilibrium.

The results include: set-aside areas, crop areas, animal numbers, costs, and farming incomes (GVA at market prices, plus direct payments) compatible with the economic accounts for agriculture at NUTS2 level; Pillar 1 budget outlays; consumer welfare indicators; and environmental indicators (N, P, K, NH_3, global warming emissions (greenhouse gases measured as CO_2 equivalents), water); as well as bilateral trade flows, prices, market quantities (at Member-State level), and intervention sales and subsidised exports.

The CAPRI simulations comprised runs to 2009 as follows:

- A reference run (Agenda 2000) based on trends and other assumptions compatible with those for other models and outlook analyses, and calibrated to 1997–99 prices and quantities.
- A MTR proposal run, in which direct payments are assumed to be uniform at regional level and based on 1997–99 (not ~2001) data, subject to dynamic and "capped" (€300,000 per farm) modulation.

Comparing these two runs, the MTR was estimated to lead to reduced supplies of crops (e.g. cereals by 7.4%) and red meat (beef by 6.6%) but price rises (cereals 0.6%, beef 5.6%). Falls in (a) farm output value (1.3%), (b) FEOGA budget outlays (8.9%), (c) farming income (0.14%), and (d) consumer welfare (6.4%) lead to a rise of 0.08% in overall net EU-15 welfare ((c) + (d) − (b)). Environmental effects are positive; global warming potential reducing by 5%, due to a drop in production of cereals coupled with an expansion of set-aside and fallow land, and a reduction in cattle production, and N surpluses by 3.4%. Structural effects on farm size are uncertain. Decoupling of premiums increases allocation efficiency and may speed up farm size growth. On the other hand, certain parts of the proposal such as farm-specific premium ceilings, and the exclusion of small farms from modulation and farm audits, may countervail this.

The Commission-published account of the CAPRI results contains little territorial commentary. Results in Britz (2004) show major decreases (at least 13%) in set-aside and fallow land in Wales, parts of Ireland, southern Finland, and parts of Greece and Austria, with major increases (over 3%) in many parts of Spain, Portugal, southern France, and Greece. There are also major decreases (at least 10%) in total premium payments per hectare in parts of central France, north and south Italy, east England and southern Greece, and major increases (−2% up to 37%) in most other regions except in Germany (where the drop is "around − 9%"). Global warming emissions drop most in Spain, central and southern England, southern France and in parts of Austria and Greece.

The drop in cereal supply is said to be "rather pronounced in regions with very low yields and [a] high share of direct payments in income for the reference run", e.g. durum wheat area in Portugal down by 60%. Total premium payments are estimated to rise in regions with high levels of permanent grassland and cattle production (with consequent rises in land rent levels, so that income distribution effects depend on land ownership). It is argued that "uniform premiums at Member State level would provoke a redistribution from more productive regions … to less productive ones". Also, "[r]egions specialising in beef production often receive higher premiums per hectare of grassland than per hectare of arable land from the COP [cereals, oilseeds, protein crops] scheme. In those regions, an identical premium shifts support towards arable crops."

In order to study the relationship between CAPRI impact measures and the EU's social and economic cohesion objectives, the CAPRI results were first apportioned from NUTS2 to NUTS3 using the method described in Section 4.2.1. These results were then analysed using mapping and linear regression techniques.

Three CAPRI measures of policy impact (both differences between 2009 estimates for MTR proposal implementation and those for the reference scenario, i.e. absence of MTR CAP reform) were considered in this analysis: CAP direct (premium) payments, farm income calculated as Gross Value Added (GVA) plus CAP premium payments, and global warming potential (expressed in terms of CO_2 equivalents).

The following trio of maps show these variables expressed as percentage changes from the reference level. Under MTR reform, CAP payments (Map. 6.1) change by more than about 25% in relatively few regions, such as the Low Countries and parts of northern Germany and northern Italy (increases) and southern France and Austria (reductions). Farm incomes (Map 6.2) are only marginally affected, with changes of more than 5% apparent only in a small number of NUTS3 regions in France (mainly in the south) and Austria (both show falling incomes) and in some or all of Northern Ireland, Belgium, northern Italy, Denmark and Sweden (all show rising incomes). Of course, these percentage changes reflect the relative size of the MTR effects and the level of farm income in the base period (1997–99). As regards CO_2-equivalent emissions, Map 6.3 shows that most regions were expected to experience a very slight reduction in CO_2-equivalent emissions (of between 0 and –1%) as a result of the MTR proposals. The only regions experiencing small increases (less than +1.5%) are mid-Sweden, south-eastern Italy and a small part of the Netherlands.

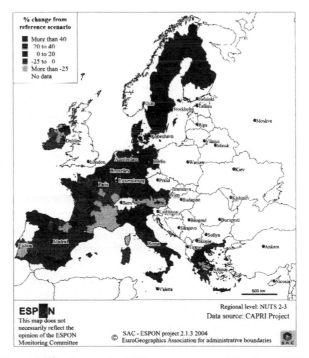

Map 6.1: Percentage change in CAP payments resulting from MTR proposals compared to the reference scenario

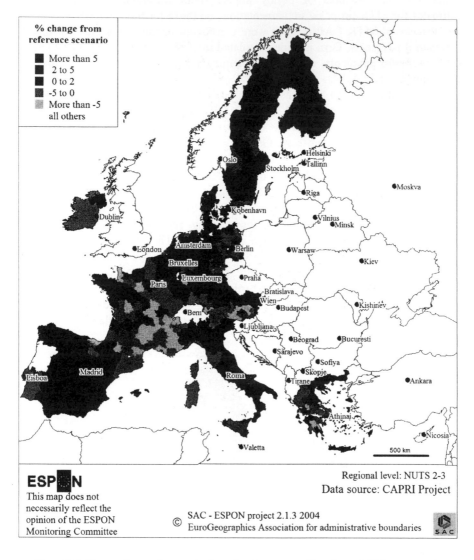

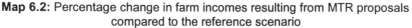

Map 6.2: Percentage change in farm incomes resulting from MTR proposals
compared to the reference scenario

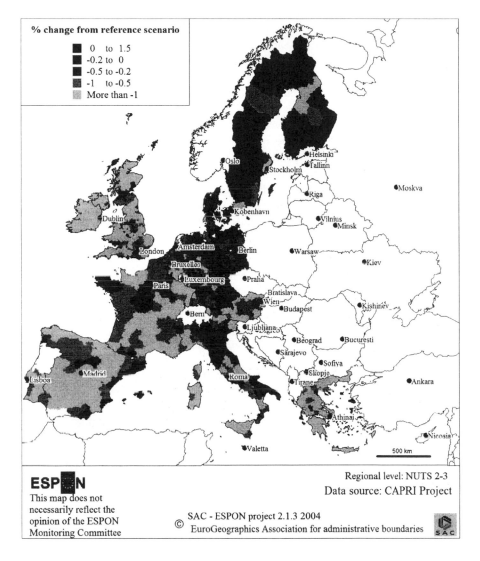

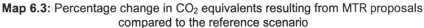

Map 6.3: Percentage change in CO_2 equivalents resulting from MTR proposals compared to the reference scenario

In terms of statistical analysis, the relationship between the percentage change in each of the CAPRI variables and cohesion indicators was explored. The 3 cohesion indicators were the same as those used in Chapter 4, i.e. GDP per inhabitant and the unemployment rate, both in 1999, and the population changes between 1995 and 1998. The results (see Table 6.1) suggest that the MTR CAP reform proposals would have increased CAP direct payments more in those NUTS3 regions with higher values of GDP per inhabitant, i.e. the generally more prosperous areas. However, there was no statistically significant relationship with NUTS3 unemployment rates. The results also indicate a negative relationship between the difference in CAP premiums and increases in population between 1995 and 1998. Thus, as a result of MTR implementation, CAP premiums would have increased more, compared to the benchmark scenario, in those areas with more slowly growing populations in the late 1990s.

Regressions using farm GVA plus CAP direct payments showed no statistically significant relationship with any of the cohesion indicators, suggesting that the overall impact of the MTR proposals on farm incomes would be territorially neutral.

Table 6.1: Pearson correlation coefficients between the estimated impact of MTR proposals and socio-economic indicators for NUTS3 regions

Impact variables	GDP per head	Unemployment rate	Population change 1989–99
Direct Premium payments	0.164(**)	—0.024	—0.069(*)
N	890	831	821
GVA plus CAP92 Premiums	−0.022	−0.057	−0.006
N	892	833	823
Global warming potential	−0.204(**)	−0.062	−0.048
N	1061	971	883

* and ** indicate significance at the 0.05 and 0.01 level, respectively, using a 2-tailed test.

Finally, the environmental indicator, global warming potential (represented by percentage change in CO_2-equivalent emissions), was positively correlated with GDP per head, suggesting that the greatest percentage increases in emissions would tend to occur in the more wealthy areas of Europe.

Table 6.2 shows the results of equivalent analysis, but in this case considering the relationships between the impact of the MTR proposals and accessibility indicators rather than cohesion indicators.

Table 6.2: Pearson correlation coefficients between CAPRI-estimated impacts of MTR proposals and accessibility indicators

	Direct payments	GVA plus CAP92 premiums	Global warming potential	Accessibility indicators[1]		
				Micro	Meso	Macro
Direct payments	1	0.099**	0.101**	−0.198**	−0.194**	−0.079*
N	890	890	889	883	886	886
GVA plus CAP92 premiums	0.099**	1	−0.060**	−0.120**	−0.052	0.022
N	890	892	889	885	888	888
Global warming potential	0.101**	−0.060**	1	−0.242**	−0.292**	−0.292**
N	889	889	1061	1050	1053	1053

* and ** indicate significance at the 0.10 and 0.05 level, respectively, using a 2-tailed test.
[1] The lower the value of the indicators, the greater the accessibility of the region.

At all three spatial scales, the largest positive impacts on direct payments tend towards the more accessible regions. Similarly, the greatest percentage increases in farm incomes (GVA plus CAP92 premiums) tend to be associated with more accessible regions at the micro level while no significant correlations were found at the meso and macro level. The results for global warming potential mirror these findings, with the largest percentage increases in CO_2 emissions tending towards the more accessible regions of the EU, in this case at the local, meso and macro levels. However, it should be borne in mind that the magnitude of estimated changes in all three variables is small.

Similar analysis was undertaken with the OECD regional typology of 6 main categories of NUTS3 regions, i.e. predominantly rural and leading, predominantly rural and lagging, intermediate and leading, intermediate and lagging, predominantly urban and leading, and predominantly urban and lagging. Five dummy variables represented these categories, using Type 6 (predominantly urban and lagging regions) as the reference area type.

The results (Table 6.3) indicate that, as a result of the MTR implementation, CAP payments would, compared to the predominantly urban and lagging areas, decrease in rural and leading areas, and probably in both types of intermediate regions. Levels of farm GVA plus CAP premiums would fall, as a result of MTR implementation, in all OECD types except predominantly urban and lagging areas, and by approximately the same amounts (not shown).

Table 6.3: Regression analysis of CAPRI-estimated impacts of MTR proposals on OECD area categories

	Predominantly Rural		Intermediate		Predominantly Urban	
	Leading	Lagging	Leading	Lagging	Leading	Lagging
Direct Premium Payments	0.447**	−0.171	0.376*	0.367*	−0.288	Ref.
Farm GVA plus Direct Premium Payments	0.033**	0.026**	0.037**	0.032**	0.027**	Ref.

* and ** indicate significance at the 0.10 and 0.05 levels, respectively, using a 2-tailed test.
"Ref." indicates the type of area with which the others are compared.

6.4. CAP/RDP reform in the New Member States (NMSs)

6.4.1   Introduction

Analysis of the impact of CAP/RDP reform in the NMSs is complicated by a number of factors (IAMO, 2004), including:

- The broad socio-economic transition towards a mature democracy and a market-based economy, with a major influence on NMS living standards
- The effects of EU accession itself, including free east–west trade within the EU-25 as well as domestic effects
- The precise way in which CAP reform will apply to these countries, following long and strenuous negotiations over the adoption of the "old" (Agenda 2000) CAP
- The considerable pre-accession aid delivered to these countries by means such as the SAPARD fund (see Chapter 5), the pre-accession Europe Agreements on trade, and the growing commercial anticipation of successful enlargement in 2004.

In general, the following general tendencies in the NMS land sector are expected to occur after EU accession:

- Limiting regulations and restrictions on the purchase of land will have to be abolished.
- In order to consolidate and enlarge the competitive and intensive core of their farm sectors, the NMSs will have to adopt land legislation much more favourable to tenant farmers.
- Access to the EU's system of direct aid will increase farm incomes, and therefore land prices and rents.

- The institutional and political convergence will activate and enlarge the markets for land in the candidate countries, while gradually integrating them into those of the EU-15.

The other effect of integrating the land markets of the NMSs, where effective land values are currently about 5–20% of the Community average, will be a considerable increase in land prices, in particular for the major crops. This will further decrease the comparative competitiveness in the NMSs, and moderate forecasts of cereal and oilseed crop surpluses. The rising cost of fodder also strengthens the forecast of under-competitiveness in the livestock sector (Pouliquen, 2001, p. 67). From the spatial viewpoint, the differences between more-favourable and less-favourable areas might be expressed particularly strongly.

Table 6.4 shows budgetary estimates of the costs of CAP adoption by the NMSs under the agreed enlargement conditions, i.e. before the June 2003 CAP reform agreement, which did not directly involve NMS ministers. The commitments for rural development are substantial.

Table 6.4: Estimated CAP expenditures, indicative allocations and RDP commitments to accession countries, 2004–2006 (million €, 1999 prices)

	Czech Rep.	Estonia	Hungary	Latvia	Lithuania	Poland	Slovakia	Slovenia	Cyprus	Malta	Total
Total Direct Payments											
2005	169	17	265	25	68	557	73	27	9	0.1	1,211
2006	204	22	316	31	84	675	88	33	11	0.3	1,464
Market Support (Budget Heading 1a)											
2004	45.0	13.6	63.6	8.9	23.2	135.2	16.9	14.9	4.9	0.7	327
2005	109.0	33.4	151.9	21.6	56.1	349.8	48.1	38.3	11.8	1.71	822
2006	111.0	34.4	152.0	23.6	59.2	376.5	49.2	38.8	11.5	1.7	858
Rural Development Commitments											
2004	147.9	41.0	164.2	89.4	133.4	781.2	108.2	76.7	20.3	7.3	1,570
2005	161.6	44.8	179.4	97.7	145.7	853.6	118.3	83.9	22.2	8.0	1,715
2006	172.0	47.7	190.8	103.9	155.1	908.2	125.8	89.2	23.9	8.5	1,825

In the run-up to accession, the application of direct payments has been controversially discussed within the EU-15, within the NMSs, and with the Commission. The core argument in favour of the straightforward application of these payments in the NMSs as in the EU-15 is that they are part of the CAP *acquis*, and the permanent exclusion of the NMSs from direct payments would not reflect the EU principle of a single market for agricultural products (EC,

2002a, p. 5). On the other hand, the application of direct payments without adaptation could have some counterproductive side-effects (EC, 2002a, p. 5), including negative impacts on restructuring, and considerable income disparities and social distortions in the rural societies of the new Member States. These impacts might well create imbalances both within rural areas (due to wide differences in land ownership) and between rural and urban areas, without adequately addressing the requirements of semi-subsistence farms. However, many of the arguments against the application of direct payments in the NMSs can also be used in the EU-15. A principal problem is that direct payments do not help small-scale and semi-subsistence farming, because this sub-sector has no significant resource base.

The conclusion, reached at the Copenhagen summit in December 2002, was to start direct payments at a low level, combined with intensified support for restructuring, in particular through rural development actions. Direct aids for the new Member States will be phased in over 10 years (Table 6.5), from 25% of the full EU rate in 2004, to 100% by the year 2013. Furthermore, there is the possibility of co-financed top-up direct payments.

The new Member States will have the option to grant direct payments during a limited period in the form of a decoupled area payment applied to the whole utilised agricultural area. On the basis of its total envelope of direct aids and its utilised agricultural area, an average area payment would be calculated for each country (EC, 2002d, p. 4). The selection of the implementation model of the direct payments (full or simplified schemes, and modulation) will have a decisive impact on the effects, including spatial variation of the prospects for the development of the agricultural sector. In addition, a package of rural development measures will be available.

Table 6.5: Phasing-in of direct payment rates, 2004–2013, and budgetary outlays, 2004 to 2006, in the NMSs

	Basic direct payments	Top-up direct payments	Outlays (million €, 1999 prices)
2004	25%	30%	1211
2005	30%	60%	1464
2006	35%	65%	1743
2007	40%	30%	
2008	50%	30%	
2009	60%	30%	
2010	70%	30%	
2011	80%	20%	
2012	90%	10%	
2013	100%		

6.4.2   Commodity and spatial effects

There are several studies which forecast scenarios after entry of the NMSs (EC, 2002a, 2003a), most agreeing on a decline in livestock production and a modest growth in cereal and oilseed production (Pouliquen, 2001). The main effects of the application of EU price policy in the candidate countries will be to encourage cereal production (due to rising price level) and discourage feed consumption. The effects on beef and dairy production are slightly positive, but not enough to cause a significant increase compared to current production levels. Pork production is likely to decline, at the same time as its consumption increases. The major impact of direct payments on production would be a further shift towards coarse grains and a faster development of specialised beef production, subject to the suckler cow premium ceilings (EC, 2002a, p. 3).

The assumption of rising price levels (often "translated" into rising land-use and ensuing spatial effects) depends on the appreciation of the real exchange rate, and the future alignment (or otherwise) between Community and world prices. A continuing convergence of farm prices towards levels of the Union has been observed, and will have a direct impact on the competitiveness of NMS farm prices (Pouliquen, 2001, p. 15). There are even examples where farm prices in NMSs have attained the level of EU-15 countries, and cannot increase much further.

Baldock and Tar (2002) estimate that the area of cereals seems likely to expand, partly absorbing land currently sown and altering it to crops where contraction may occur, such as potatoes. Some abandoned arable land might come back into production. Intensification seems unlikely to reach EU levels, because of the lower land and labour prices, moreover the limited availability of credit and the low inputs of pesticides and inorganic fertilisers. More marginal cereal land may continue to justify little investment or new infrastructure and may gradually be abandoned or be converted to other uses, e.g. forestry.

A more differentiated assessment of the CAP impacts is expected for livestock production. Semi-subsistence farming is important in the dairy sector, and with large numbers of small producers, many observers expect them to become less competitive and to withdraw from farming in sizeable numbers. This could be accelerated by the health and hygiene standards within the EU and the introduction of the milk quota regime. Therefore dual effects are probable: abandonment from small plots of semi-subsistence farmers and intensification of the large producers.

These views are shared by more recent assessment studies stating that husbandry will decline. An impact assessment of the CAP reform proposals (which incorporates the effect of decoupling direct aids) forecasts for EU-25 that the utilised agricultural area will rise stronger than the resulting yields: the new Member States add about 38.0 million hectares of UAA to the 130 million hectares of the old Member States, representing an increase of 30%. The EU-25 would produce in 2006 about 30% more cereals with 42% more cereal area and 25% more oilseeds with 37% more oilseed area (EC, 2003a, p. 12).

6.4.3   Rural development policies in the NMSs

There is a broad consensus that rural development policies should be territorially defined and based on an integrated approach, embracing the economic, social and environmental aspects of rural development (OECD, 1997, p. 22). The large portion of the accession countries dominated by rural areas underpins the relevance of rural development programmes.

Rural regions in the enlargement area are affected especially by transformation problems. They show sharp economic spatial disparities and have few urban centres. To a certain extent, the mix of sharp declines in production and employment levels, poor infrastructure and poor transport accessibility could lead to a massive wave of out-migration from rural regions, and as a consequence, to the collapse of their socio-economic viability (EC, 1999, p. 50). Yet, in many NMSs the formulation of rural development policies is at a rather early stage and they are still mainly targeted at the agricultural sector and the basic rural infrastructure (OECD, 1997, p. 22).

A tradition for spatial development and regional policies similar to those of many EU-states and as defined in the EU Structural Funds hardly exists. This can be seen through the lack of spatial development and regional policy instruments and institutions as well as by the fact that, in general, independent regional levels in the political and administrative territorial system do not exist (EC, 1999: pp. 48-49). The OECD proposes an effective, well-designed and suitably targeted institutional adjustment, which is crucial for rural development policies.

> Given their territorial and multi-sectoral character, rural development policies and programmes involve a wide array of actors including sectoral ministries, government agencies, intermediate and local administrations, local private business, trade associations and voluntary organisations. Therefore, an institution needs to be designed with the responsibility and authority to lead and coordinate rural development policies (OECD, 1997, p. 126).

As outlined above, most candidate countries cannot be expected to develop the administratively more demanding rural development measures on the basis of their current limited administrative capacity and experience alone (Baldock and Tar, 2002, p. 12). The two programmes agreed at the European Council meeting in Berlin as part of the Agenda 2000 proposals, the Instrument for Structural Policies for Pre-Accession (ISPA) and the Special Action for Pre-Accession Measures for Agriculture and Rural Development (SAPARD), were aimed particularly at this lack of institution building and supported the (technical) implementation of territorial development policies in the applicant countries.

As an outcome of the Copenhagen summit in December 2002, the EU decided on a package of rural development measures in accession countries, including:

- Early retirement of farmers
- Support for less-favoured areas or areas with environmental restrictions
- Agri-environmental programmes
- Afforestation of agricultural land
- Specific measures for semi-subsistence farms
- Setting up of producer groups
- Technical assistance
- Special aid to meet EU standards.

Additional rural development measures (investment in agricultural holdings, aid for young farmers, training, other forestry measures, improvement of processing and marketing, adaptation and development of rural areas) will be financed from the Structural Funds (EAGGF Guidance Section; EC, 2002d, p. 2).

6.4.4   Distinction between NMSs

The pre-conditions for the NMS accession with respect to implementation of rural development policies are highly diverse. For example, Slovenia and Slovakia recently established central coordinating bodies responsible for rural development policies. However, financial support for rural areas is still strongly linked to agricultural production, while support to other activities in rural areas is still negligible. In Latvia, Lithuania, Albania and Bulgaria, the restructuring of agriculture still predominates, and minor attention is being paid to the specific problems of rural areas at present. However, a growing number of projects aimed at tackling specific rural problems are being undertaken, usually with the backing of international donors. In Romania, no institutions exist for the specific purpose of promoting rural development, and rural policy is still equated with agricultural policy and primarily a centralised approach (OECD, 1997, p. 22).

This leads to the two main types of institutional development. In a first group of countries, central institutions coordinating rural development policies have been created, and the sector approach is integrated into regional strategies. In the Czech Republic, Estonia, Hungary, Poland, Slovakia and Slovenia, rural policies are distinct from agricultural policies; decentralisation of the decision-making process and the involvement of local actors in solving community problems seems to be most advanced, and central coordination bodies responsible for rural development policies are already established. Whereas rural development policies are conceived in close relation to regional policies, the financial support for rural areas still remains heavily concentrated on agricultural production, and support to non-agricultural activities in rural areas is marginal (OECD, 1997, p. 130). Expectations of a more integrated approach and multi-sectoral programmes were disappointed by the practice of the pre-accession programmes (e.g. SAPARD) which tightly focused on competitiveness and sector aspects.

Latvia, Lithuania, Albania, Bulgaria and Romania are the second group. The discussion of the specific problems affecting rural areas is at an early stage and the restructuring of agriculture still predominates. A growing number of individual projects are aimed at solving specific rural problems, but there is no separate institution which coordinates rural development issues.

There is thus a great challenge for applying strategies for integrated rural development policy in the NMSs. The following aspects are particularly relevant and require greater attention (OECD, 1997, pp. 131–132):

- The low level of economic development leads to a domination of macroeconomic strategies, and preferential treatment of restructuring and privatisation issues. Programmes to decrease spatial divergence and support non-urban areas have still to be started. However, the tight central budgets leave only little resources for financing local governments' development initiatives.
- Public funds are channelled to rural areas almost exclusively through agricultural policy measures, partly because of the still important role played by agriculture and strong agriculture lobbies.
- Rural development is the result of a long-term process of institutional evolution and socio-economic development.
- The disruption of local development, and the experience of collectivisation and artificial creation of rural settlements (agro-centres) and the recent transformation period have increased insecurity and threats for local population. In many rural areas this has contributed to the weakening of the identification of the rural population with the area in which they live. There is a rising need to re-shape local identities and to nurture potential of rural areas in NMSs.
- Insufficient means at local level to solve local problems, the dominance of top-down approaches, and lack of secondary and vocational education available for the rural population are further important handicaps for rural development activities.

6.4.5   Mountain areas in the NMSs

The spatial importance of mountain and other less-favoured areas results not only from their extensive utilised agricultural areas but also from the economies of spatial cohesion, which may offer opportunities for a sustainable future. As a specific spatial category, mountain areas reveal features of spatial divergence and environmental impacts more clearly than other areas. The case of the mountain areas is therefore presented here in order to focus on the particular needs of less-favoured areas in the accession countries. Appropriate strategies for agricultural and rural development policies will play a decisive role, particularly in the more marginalized areas of the NMSs (Dax, 2001, p. 2).

Common aspects for all mountain regions are:

- Widespread poverty, with weak economic integration and participation by the mountain population in the economic and social life of the country.
- Restricted access to public services such as hospitals, primary schools, or cultural activities.
- Lack of infrastructure, such as roads, telecommunications or electricity, while systems that once functioned, such as irrigation, sewage and heating networks, have sometimes fallen into ruin.
- Dramatic levels of open and hidden unemployment, even if statistics are lacking.
- Significant out-migration, although, where family and social structures have remained more intact in the mountains, and semi-subsistence holdings may offer a modest living standard, some areas have had substantial population growth due to in-migration from urban centres.

Many parts of the NMSs are characterised to a large degree by such regions, for instance the Carpathian mountains which extend over parts of the Czech Republic, Slovakia, Hungary, Poland, Ukraine and Romania. Due to natural difficulties and problems of the restructuring of agriculture, the income potential from agricultural production in these areas is substantially lower than in lowland areas. These areas are furthermore threatened by trends of growing interregional disparities. Nevertheless, the features of mountainous regions in the NMSs are very diverse (Dax 2001, p. 6). Whereas Slovenia and Poland have mostly well-developed infrastructure in the mountain regions, other countries have limited social problems in mountains and rather good infrastructure (e.g., the Czech Republic, Slovakia and partly Bulgaria). Still others, like Romania and Albania, have a lot of small private farms, suffer from overpopulation and a lack of job possibilities, which implies high unemployment rates and badly developed infrastructures.

The degree of implementation of a specific LFA framework and policies in NMS mountain areas is highly different. In the run-up to accession, some countries (Czech Republic, Poland, Slovakia and Slovenia) already provided support to farming in marginal areas and in particular mountain areas, especially to grassland-based farming methods. Among the three Baltic countries, only Lithuania has established a similar programme to date. On the other hand, Bulgaria, Estonia, Latvia and Romania have not yet developed less-favoured area type schemes (Dax, 2001, p. 6). In all countries, there is a lack of an integrated approach in the mountain areas.

6.4.6   Conclusions

The CAP reform proposals are expected to generate a sustainable improvement in the medium-term perspectives of the agricultural sector of the EU-25. In the New Member States, CAP reform could secure income gains, generated by enlargement, which could reach up to 45% when taking account of the phasing-in of direct payments and rural development measures (EC, 2003a, p. 4). Although the territorially differentiated effects of the reform are rather difficult to calculate, it is concluded that the reform proposals would have diverging impacts across regions and the various sectors, leading to declines in the milk and (food) oilseed sectors, broadly stable development in the cereal sector, and significant gains for the meat sector. There may, however, be additional spatial aspects for the internal development of agricultural production in the NMSs, depending on structural and region-specific factors.

Chapter 7

7. The CAP/RDP in the Context of EU Spatial Policy

7.1. Introduction

The European Spatial Development Perspective (ESDP) and subsequent reports on Economic and Social Cohesion published by the European Commission have sought to promote a more integrated approach to policies for rural and urban areas. As noted in Chapter 3, European spatial policy is guided by three fundamental goals of the European Union:

- economic and social cohesion
- conservation of natural resources and cultural heritage
- more balanced competitiveness of the European territory.

EU spatial development policies seek to promote sustainable development of the EU in accordance with the following policy guidelines:

- Development of a balanced and polycentric urban system and a new urban-rural relationship,
- Securing parity of access to infrastructure and knowledge, and
- Sustainable development, prudent management and protection of nature and cultural heritage (EC, 1999, p. 11).

The key questions to be considered in assessing the relationship between agriculture and rural development policy on the one hand, and EU spatial policy on the other, include the following:

- To what extent are the objectives and instruments of the CAP and RDP compatible with the concepts of balanced polycentric urban development and new urban-rural partnerships?
- To what extent are the outcomes of CAP and RDP measures in conformity with the EU cohesion objectives?
- To what extent do the CAP and RDP instruments support sustainable development, prudent management and protection of nature and cultural heritage? and
- To what extent are EU measures to promote parity of access to infrastructure and knowledge compatible with the CAP and RDP?

These questions have been addressed in this report largely through the evidence presented in Chapters 4, 5 and 6. In the following sections, the nature of the concepts guiding EU spatial policy is further explored and the empirical findings are then related to each of these more explicitly.

7.2. Balanced competitiveness, polycentric urban development and new rural urban relationships

7.2.1   CAP and Agenda 2000 reform

The Agenda 2000 reform provided a new framework for rural development policy, the Rural Development Regulation (Reg. 1257/99), including: principles of multifunctionality of agriculture; multisectoral and integrated approach to the rural economy; flexible aids for rural development, based on subsidiarity and promoting decentralisation; and transparency in the drawing up of, and management processes for, rural development plans. Thus, the preconditions for a more endogenous development seem to be strengthened through these principles. The Regulation offers some new scope for governments to tailor measures more effectively to meet the varied local needs of rural areas, at least from the conceptual level for programming.

However, the menu and mode of delivery of Pillar 1 measures, which consume the vast majority of the total EU agricultural budget relative to Pillar 2, provide little incentive for promotion of an integrated multi-sectoral endogenous approach to the development of rural areas. The mono-sectoral approach is indicative of a traditional perspective on rural-urban relations, where farmers were supported to provide food for the expanding urban populations. From this perspective it may be argued that the Pillar 1 component of the CAP is not consistent with the goal of balanced regional and rural development.

Article 33 measures in the RDR provide countries with instruments to increase the scope of action of farmers and people in rural areas. However, on average only about 10% of funds of the RDP are foreseen for these measures in the EU. In addition, most of these measures are only eligible for the farm sectors (Dax, 2002b). Budget constraints are also expected for the overall structure of Pillar 2 measures. In most countries, the stakeholders believe and first assessments reveal (Dwyer *et al.*, 2003) that the budget is much too small to adequately deliver the programme objectives in the period 2000–2006 and that a substantial increase will be required for the next programme period, starting in 2007. This perspective was shared by the European Commission assessment in the Mid-Term Review (EC, 2002b). However, subsequent changes to the reform proposals resulted in very modest modulation effects from Pillar 1 to Pillar 2, and the results will probably be less relevant than would appear from the prominent place in the discourse of (past and) current CAP reform (Baldock, 2003, p. 100).

7.2.2   Networking and cooperation

Encouragement of interaction and cooperation between neighbouring cities and towns and their surrounding rural areas is essential for developing polycentrism in a region. It is therefore important to examine the extent to which the CAP and RDP promote territorial-based networking and cooperation to enhance competitiveness. But "cooperation is a delicate flower" that is not easily introduced and that can only survive over the long run if the distribution of benefits bears some perceived and acceptable relationship to the distribution of costs between partners, an outcome sometimes difficult to achieve (Parr, 2003, p. 15). There are at least two factors which condition the extent of cooperation. One is that some kind of identification of the citizen, the householder, the worker, the manager or the firm with the region (territory) within which the cooperation should take place, while the other is that the structure of the local government should foster cooperation.

As an integrated rural development strategy, LEADER may provide a response to the need for promoting cooperation between rural areas, and between urban and rural areas. LEADER allows experiments with local (territory-based) small-scale actions (pilot projects) using the endogenous potential of the area. The bottom-up approach allows the local community and the local players to express their views and to help define the development course for their areas in line with their own views and plans. LEADER is implemented by local action groups (LAGs) that are organised on the basis of partnerships to facilitate cooperation between different actors in order to implement integrated multi-sectoral programmes that require linkages between several sectors of activity so that rural innovation programmes can be more coherent (Van Depoele, 2003, p. 49f). This and many other positive assessments of the LEADER approach reveals the potential provided through this Community Initiative. However, most evaluations also emphasise the need for gradual and long-term commitment that takes account of the limits of local actors' participation, the difficulties of cooperation at local and regional levels, the sectoral compartmentalisation of many regional contexts, and the still limited experience on exchanges. Nevertheless, the LEADER programme has been a significant catalyst for innovative approaches in rural areas, and could serve as a model for more comprehensive rural development. This is considered further in Chapter 8.

Projects supported within LEADER may therefore support strategies using a polycentric development model by enhancing awareness of regional potentials and facilitating cooperation and networking between different actors of the agricultural sectors and beyond it. This implies close cooperation with other (European or regional) programmes which can supplement the networking of actors and increase regional effects. From the territorial perspective of rural development, many Structural Funds programmes (INTERREG, some EQUAL; Objectives 1 and 2) and environmental programmes (LIFE, NATURA 2000;

Local Agenda 21) are quite relevant and provide examples of pilot actions in rural areas.

It is disappointing that only a few Rural Development Programmes have taken up the option to develop integrated programmes (or to partly integrate some of their measures). The weak application of this principle seems primarily due to the institutional framework of RDP within agricultural policy and the administrative structure which favours the continuation of existing (agricultural) measures within RDP (Mantino, 2003).

Polycentric development promotes the enhancement of the accessibility of urban and rural areas through better infrastructure on the one hand, and the improved assignment of functional tasks of urban-rural relationships on the other. As most measures of CAP and the RDP are conceived horizontally and encompass all the agricultural area of the countries, there is hardly any focus on geographical differentiation or assessment of the impacts of infrastructure development on the sector. Accessibility is split into different aspects and has to be analysed on the targets to be accessed. For large parts of the programmes, there is a particular lack in coherence between RDR funds and other EU policies and funds. The dominant picture is also one of relatively weak integration between measures, and between these regulations and other national and regional rural funding. The continuing preoccupation of many RDR programmes with agriculture and the very restricted discussion of rural area problems recall the need to widen the scope of measures and address these concerns in future programmes. An integration approach would inevitably require discussions in the process of plan development of how to assign the functional tasks of urban and rural areas, and how to deal with these tasks in the proposed programme measures. It would be important to view the chances of rural areas in the framework of a re-designed spatial development policy. With over half of the population in the 25 Member States of the European Union (EU) living in rural areas, which cover 90% of the territory (EC, 2004c), this additional perspective to the polycentricity concept is of considerable relevance for spatial and cohesion policy.

7.3. CAP/RDP measures and EU cohesion

7.3.1   Agricultural policy and cohesion

The principal instruments of the CAP prior to the 1992 reforms, namely market support and protection measures, were designed to achieve multiple objectives including an expansion of agricultural output. Combined with technological advances, the CAP measures contributed to increasing intensification, specialisation and concentration. The spatial distribution of the incidence of market support payments has been linked to the intensity of farming and the extent to which different farm enterprises attract support payments. Variations in the intensity and scale of farming operations are influenced by many factors which are not distributed uniformly across the regions; rather they frequently combine, resulting in some regions having distinctive sources of comparative advantage for specific types of agricultural production. The trend towards increasing specialisation when combined with regional differences in comparative advantage for particular farming types has resulted in an increased level of regional concentration of production.

In broad terms the CAP has contributed to improving the economic and social situation in rural Europe. CAP support mechanisms have helped to maintain agricultural production in some regions at levels that would not have been possible in an environment of more open competition. The specific instruments to ensure guaranteed prices and to provide protection from lower-priced imports have enabled more farms to survive than might have happened in the absence of the policy. This is especially the case in some weaker rural areas where opportunities for alternative forms of economic development are more restricted. The Guidance Section of the EAGGF has provided assistance for structural reform and modernisation of on-farm production and off-farm processing of farm output.

The role of the CAP in supporting the rural economy has since the late 1980s been complemented by the consolidated and enlarged structural funds. In particular assistance towards investments in physical infrastructure, water distribution systems, farmyard facilities for storing and managing sources of pollution, and also investments in targeted training and advisory programmes have complemented the objectives of the CAP and more recently those of the rural development programmes.

The positive outcomes noted above, however, should not be allowed to disguise some serious concerns that have arisen from the application of the CAP in different parts of the EU. The empirical evidence in relation to the spatial distribution of the incidence of market price supports demonstrates that the highest levels of payments per AWU and per ha UAA tend to occur in some of the richer regions of the EU. Overall, the incidence of price supports is lowest in the poorest regions due to weaker agricultural structures, and also in regions with the highest unemployment rates. This outcome is at variance with the economic and social cohesion objectives of the EU.

Since the early 1990s there have been a number of initiatives to re-orientate the CAP towards international market conditions. The shift towards greater emphasis on supply control measures, compensation payments and more comprehensive rural development programmes has the potential to significantly alter in a positive way the relationship between the CAP and cohesion objectives.

However, the introduction of compensation via direct payments is problematic for two reasons (Buckwell, 1996): first the level of payments is not sufficiently linked to the income reductions associated with the lowering of commodity price supports which has led to overcompensation of some groups of farmers especially cereal growers who are mostly located in some of the richest EU regions, and second there has not been a clearly articulated rationale to support an indefinite continuation of such payments for a once-off policy change.

Extending the provision of such payments to the new Member States will require significant adjustments in order to avoid further market distortions and increased levels of social inequality between the farming and non-farming populations.

The imposition of ceilings on compensation payments has ameliorated to some extent the effect of variations in the intensity and scale of farming. Thus the most recent evidence for the distribution of direct payments at NUTS3 level shows a significant degree of consistency or complementarity with the Structural Fund Objectives in relation to per capita GDP and unemployment rates and, therefore with the cohesion objectives. The contribution of direct payments to total agricultural income is particularly strong in low-intensity farming regions including upland areas where cattle and sheep farming systems are the most common types of farm enterprises. They are also significant in some regions with large areas under cereals which include some relatively underdeveloped sub-regions as, for example, in Spain.

A number of issues need to be considered in relation to the role of such payments in the future, including the likely level of public support for their continuation over the longer term; the relative rate of economic return from such payments; whether they hinder or restrict diversification; and their impacts on the wider economy especially the agri-processing sector. For example, the move towards decoupled direct payments may lead to reduced numbers of livestock that may in turn result in lower levels of purchased inputs and also less volume for processing, in which case there may be further rationalisation as processors compete more for supplies. The impacts of such off-farm adjustments are most likely to be greater in rural areas and thus may make the challenge of territorial cohesion more difficult to achieve.

7.3.2   Rural development policy and cohesion

Progress towards establishing a comprehensive rural development policy with a stronger territorial dimension has been very slow. The introduction of the Less-Favoured Areas scheme in the 1970s was the first explicit recognition of the need for special assistance in designated areas and as such was an important first step in the process of introducing a territorial dimension into a mainly sectoral policy. Despite the significant conceptual and methodological difficulties associated with measuring the impact of LFA payments it is likely that the overall outcome complements the economic and social cohesion objectives. However, any conclusion regarding the impact of LFA payments must be qualified by the following concerns:

- the small share of the EAGGF expenditure allocated to LFA payments given the severity of the problems to be overcome in these areas;
- the intra-regional distribution of LFA expenditure is linked to volume of production; thus it does not address sources of local inequalities;
- the availability and scale of LFA supports, especially when coupled with other subsidies, has in some cases restricted progress in relation to

restructuring of production that would lead to larger and more competitive farms;

- LFA and related supports may have also hindered efforts to promote alternative land uses, especially afforestation, thereby resulting in suboptimal resource use.

The reforms of the early 1990s included the introduction of a number of accompanying measures, of which agri-environment schemes were the most notable. Such schemes have multiple objectives. As eligibility to participate tends to be contingent on relatively low-intensity farming, though not exclusively as for example in parts of Spain, the distribution of participants and levels of payment per AWU and per UAA are expected to be highest in the weaker rural regions. Therefore, such schemes are expected to contribute to the achievement of the economic and social cohesion goals. However, the statistical analysis reported in Chapter 4 refutes this hypothesis at the level of NUTS3 regions across the EU. This result may reflect a tendency for some Member States to give higher priority to agri-environment measures in response to the severity of the problems that have already arisen from their intensive production systems. Given the variation between Member States in the operating rules for agri-environment supports it is necessary to interpret cautiously any EU-level generalisations.

It is also necessary to qualify any conclusion about such payments by relating the level of expenditure to the total level of support provided under the CAP. The positive contribution of agri-environmental measures in weaker regions may not be sufficient to counter the effects of product supports in the stronger regions. Therefore, it is likely that for specific regions, total CAP expenditure continues to be at variance with the economic and social cohesion objectives. While it is difficult to empirically test this hypothesis across all the regions of the EU the evidence for Ireland in the late 1990s is supportive.

The desirability of a transition to a more elaborate framework for sustainable and integrated rural development was first discussed at the 1996 Cork Conference on Rural Development hosted by the European Commission. The ensuing Declaration, which sought to lay the basis for "making a new start in rural development policy", proved to be overly ambitious. The Commission proposals for Agenda 2000 introduced the notion of rural development as the Second Pillar of the CAP. However, by the conclusion of the CAP reform negotiations the commitment to a new approach to rural development was severely curtailed. The main outcome was the Rural Development Regulation (no. 1257/99) which falls far short of the objectives contained within the Cork Declaration. A further retreat from a vigorous and comprehensive rural development policy is evident in the Salzburg Declaration.

The RDR aims to provide a single, coherent package of support to all rural areas in three main ways:

- by creating a stronger agricultural and forestry sector,
- by improving the competitiveness of rural areas, and

- by maintaining the environment and preserving Europe's unique rural heritage.

These aims are to be achieved through the implementation of a broad range of measures which include as "accompanying measures" agri-environment schemes, LFA compensations, aid for afforestation of farmland, and early retirement aid. In addition assistance is provided for the following actions: investment in agricultural holdings, setting up of young farmers, training, improving the processing and marketing of agricultural products, and promoting the adaptation and development of rural areas. When taken as a package these measures are a positive initial step towards the more ambitious goal of sustainable and integrated rural development and ultimately the achievement of greater territorial cohesion. However, the potential effectiveness of the RDR is limited by the fact that in many countries it amounts to little more than an amalgamation of pre-existing measures to provide support for activities close to agriculture (Dwyer *et al.*, 2002). It does not provide a coherent basis for a truly integrated approach to rural development.

The evaluation by Dwyer *et al.* (2002) of the Rural Development Programmes prepared for ten countries identifies marked differences between Member States. While agri-environment schemes are a compulsory component of each RDR Programme the relative allocation of resources to this measure is highest in some of the countries with the least severe environmental problems such as Sweden, Finland, Austria and Ireland. LFA payments are of very limited importance in intensively farmed countries such as Denmark and the Netherlands in contrast to countries with extensive upland and low-intensity grazing areas as in France, Greece, Austria, Ireland and the UK. In the intensively farmed countries of Northern Europe – Denmark, Netherlands and Belgium – the emphasis has been placed on measures to improve structures and marketing.

The limited evidence available thus far in relation to the design and implementation of the RDR programmes suggests a number of significant weaknesses that need to be addressed. Fundamentally, the Second Pillar is much too closely aligned with agriculture and the imbalance in the allocation of resources between Pillars 1 and 2 is so large that it is impossible to conceive of strategies aimed at achieving the objectives of sustainable and integrated rural development which are central to the pursuit of territorial cohesion. There is an urgent need to:

- Place rural development as a component of comprehensive strategies for integrated regional development that explicitly recognise the extent of old and new types of rural-urban relations. Small-scale localised actions are not a sufficient response to the many diverse challenges confronting rural areas. In order to achieve the territorial goals associated with the Structural and Cohesion Funds a higher-level strategic approach to rural development is needed that will require closer alignment of the measures contained in RDPs with those in the Structural Fund Operational Programmes.

- Ensure that the allocation of rural development assistance attaches more emphasis to medium- and long-term development potential based on strategies to combine endogenous and exogenous resources and that extend the range of supports beyond the farming community (Terluin and Venema, 2003).

- Adjust the balance of support between CAP and RDP so that the outcomes from this policy area can become more supportive of economic and social cohesion goals. The traditional relationship between agriculture and rural development where the former is regarded as the driver needs to be reversed so that comprehensive RDPs can be regarded as frameworks for ensuring the long-term sustainability of the European model of agriculture based on the concept of multifunctionality.

- At the level of implementation there should be more objective assessment of the relative needs for rural development and more careful targeting of resources towards the elimination of market failures that work against the achievement of rural-based public goods. These include the quality of food, soil, water and atmospheric resources, biodiversity, habitats and landscape and also the development of viable and sustainable rural communities and economies. There is also a need to ensure that implementation procedures facilitate greater local participation and permit sufficient flexibility to enable local customisation of strategies. The current round of RDPs are the result of a predominantly top-down preparation process that has relied heavily on amalgamating pre-existing measures and which has maintained an approach that regards rural development as an adjunct to agriculture policy. This methodology is very much at variance with the pilot experiments involving innovative approaches to rural development promoted by the EU Commission in conjunction with local action groups with assistance from Community Initiatives such as LEADER (for a comprehensive overview of the outcomes from the LEADER approach see Moseley, 2003). The intention to mainstream LEADER-type programmes in the next round of Structural and Cohesion funding is a welcome move, especially if it is accompanied by an emphasis on the application of good practice principles as summarized by the Lukesch report (ÖIR – Managementdienste GmbH, 2004).

The challenges confronting agriculture and rural development in the new Member States are very large. The analysis by Buckwell *et al.* (1995) pointed to many risks associated with transferring the "western" CAP, even after the 1992 reforms, to the context that prevails throughout most of Eastern Europe. There are many instances of dualistic production structures, as for example in Hungary (Ferenczi, 2003), which require much greater emphasis on rural development programmes that can be integrated at the level of regions with other support programmes. In common with the rest of the EU there is a very strong case for shifting from a sectorally defined commodity support framework to a broader territorially defined set of integrated support measures which will support a stable and efficient food producing sector that will be embedded in sustainable rural environments populated by sustainable rural communities.

The commencement of the implementation in autumn 2002 of the SAPARD Regulation for pre-accession countries has provided an opportunity to support economic and social reforms in these countries. In general terms the measures supported by SAPARD mirror those included in the RDR programmes with the exception that assistance for early retirement, setting up young farmers and for producers in Less-Favoured Areas are excluded. Furthermore the assistance towards agri-environment schemes is very much restricted to pilot actions. Dwyer *et al.* (2003) noted that the impact of SAPARD to date has been limited by the relatively small amount of resources devoted to it, start up delays and an emphasis on developing institutional capacity in the central administrations. Nevertheless, SAPARD has already had a significant indirect effect in encouraging a number of accession countries (for example Hungary) to introduce as early as 1999 alternative approaches to rural development which are similar to those elsewhere in the EU. There is an on-going need for a reallocation of further resources towards rural development which must be guided by strategies containing a more explicit territorial focus and which facilitate greater participation of a wider range of stakeholders at local and regional levels. The adoption of a programme approach is a welcome initiative but this needs to be developed further with more emphasis on integration across sectors and development dimensions at regional levels, and also with more resources for local and regional capacity building. These objectives can be achieved more effectively through closer alignment of CAP and RDP policies with those for the Structural and Cohesion Funds and thus ensure a more effective approach to territorial cohesion.

7.4. CAP/RDP measures and environmental sustainability

The Sixth Environmental Action Programme places considerable emphasis on the integration of environmental policy with other policies. The Programme provides a binding framework for the period up to 2010. Already a significant number of Community environmental measures affect agricultural production and establish standards which farmers are required to meet. These are frequently supplemented by national and regional measures. These standards are almost entirely established outside the CAP framework. Once they are adopted the role of the CAP is to assist in their enforcement by facilitating and encouraging farmers to adjust their farming practices to the changing expectations of wider society in regard to the environment.

The Baldock *et al.* (2002) report has proposed a strategy for integrating agriculture and environmental policies which relies heavily on active pursuit of complementarities and synergies between the two policies. The integration of environmental concerns into the CAP requires an approach that seeks to address the broad range of agricultural production, not just individual sectors. More fundamentally agriculture policy in the future must place more emphasis on supporting a realistic alternative to the productivist model by one based on behaviours more in accordance with the principles of ecological modernisation (Hajer, 1995; Evans *et al.*, 2002). It is especially important that the levels of supports provided via agri-environment measures in New Member States are large enough to encourage a high take-up rate in order to avoid an orientation of the majority of producers towards "modernisation style" productivist patterns.

Agriculture production must respect fundamental principles such as "the Polluter Pays" and comply with international standards as laid down by Directives or Regulations relating to water, nitrates, birds and habitats, etc. Integration strategies need to be developed at national and regional levels that will enable agriculture to comply with the requirements of the Water Framework and other directives.

The task of achieving environmentally sustainable farm management practices by means of conventional support policies is a major, if not impossible, challenge. Future patterns of agriculture production will be influenced by several factors including new technological developments, adjustments in the wider economy which will impact on the availability and cost of labour, new international trade agreements, and changes in consumer preferences with more emphasis on the quality of food and also on environmental impacts of different production systems. The reforms of the CAP can assist in achieving more environmentally sustainable production systems by restricting production aids, decoupling compensation payments from production and through greater emphasis on rural development which includes agri-environment schemes as a component of Pillar 2.

Pillar 2 provides an opportunity to significantly enhance the level of integration with environmental policy, though in order to do so effectively a significantly larger share of the CAP budget needs to be allocated to the Rural Development programmes. Baldock *et al.* (2002) have identified the following possibilities for Pillar 2 measures:

- Agri-environment schemes that are designed to cater for local conditions have the potential to address a large number of environmental concerns. More attention is required in the identification of the reference levels of good farming practice, and to promoting the schemes in order to achieve higher take-up rates.
- Supports for sustainable farming in Less-Favoured Areas need to be adjusted to reflect local carrying capacities.
- Greater emphasis on targeting farm investment aid towards achieving environmental standards that may become mandatory in the future.
- More support for training, marketing and processing in order to encourage more sustainable land management and food production systems.
- They also propose several changes to the manner in which Pillar 1 measures are implemented, including:
- Amending or eliminating measures which under current circumstances provide incentives for environmentally damaging forms of production and

other environmental pressures at a level which otherwise would not arise. These include aids for tobacco, cotton, sugar and also forage maize under the arable regime.

- Breaking the link in several sectors between the extent of support provided and the volume of production. A move towards decoupling should lead to more environmentally sustainable levels of livestock production, similarly a move towards area-based payments for olive production could lead to a reduction in environmental risks.
- Making more substantive use of the principle of cross-compliance.
- Incorporating environmental considerations into marketing and food labelling policies associated with the CAP market regimes.

Horizontal measures especially in respect of organic production and training have been a feature of the programme in most Member States. However, they have been largely identified with environmentally sensitive and extensive farming areas, with the notable exception of Austria where the aim is the "ecologicalisation" of all agricultural activity.

The achievement of ESDP objectives, especially those relating to prudent management of resources, depends on effective environmental integration within the CAP. Possible mechanisms for integration include cross-compliance and a requirement of verifiable environmental standards for certain Pillar 2 measures (Baldock *et al.*, 2002). The potential benefits of an integrated approach to EU structural and regional policy instruments are also supported by the conclusions by Schramek *et al.* (1999) who recommended improved integration of existing structural and regional policy and instruments such as the LFA scheme and LEADER with agri-environmental policy.

7.5. CAP/RDP measures and peripherality/accessibility

7.5.1   CAP reform and peripherality

The pre-reform CAP protected EU producers from more competitively priced imports, and, through the intervention system for some products (e.g. beef), delayed restructuring towards high value added processing and thereby supported relatively inefficient sectors in some regions. It is very difficult to assess the impact of peripherality on agriculture under the CAP prior to 1992, since geographical patterns of farm-gate prices would have been very much a reflection of price support and intervention arrangements, and therefore probably more uniform than they might otherwise have been. Gross margins and incomes for certain livestock products were dominantly a reflection of direct subsidies, and this too would have had a tendency to mask the impact of peripherality. Certain aspects of the pre-reformed CAP discriminated in favour of the periphery: LFA subsidies, for instance, although intended for the hills and uplands rather than peripheral areas *per se*, naturally tended to benefit the more remote regions of the EU more than the accessible ones.

In these and other ways, the CAP has probably restrained some regions from achieving their full potential and left weaker regions more at risk from increased external competition which is enhanced by improvements to interregional and international transport infrastructures. Some evidence for this scenario is provided by trends in the food retail sector where technological improvements related to maintenance of food quality, accompanied by transport improvements and also by the emergence of international food retail chain stores have resulted in higher levels of competition for locally produced food products.

The removal or weakening of the link between subsidy and land use under the MTR seems likely, unintentionally, to have the effect of revealing hitherto masked disparities, as peripheral areas are able to extensify production without a loss of subsidy. It is conceivable, for instance, that the Scottish Island Areas may lose a proportion of their sheep, since for some time the cost of transporting them to the mainland by ferry has exceeded the pre-subsidy profit. It has frequently been suggested (see for instance University of Aberdeen, 2001, para 3.2.12) that one of the impacts of the MTR will be to divide farm businesses into two groups, those which are large enough to enjoy scale economies, and have sufficiently productive resources, specialising in commercial production, and those which are smaller, and which have poorer land, extensifying and diversifying; subsidies allowing them to act more as countryside custodians than commercial farm businesses. It seems likely that the former group will tend to be located in the more accessible low ground areas, whilst the latter will predominate in more peripheral areas.

Theoretical implications regarding the impact of the changing transport/travel cost environment upon the small business sector in peripheral regions seem to be consistently negative. Thus, it is anticipated that modern growth industries will be subject to increasingly strong agglomerative forces (due to the complexity of modern logistics systems, "just-in-time" delivery, and so on), whilst businesses in peripheral areas will be increasingly exposed to competition from central regions. It is within this very demanding environment that farmers (apart from the few with potential to become large-scale commercial producers) will be expected to diversify and establish new non-agricultural enterprises to supplement their income. Recent research (Copus, 2004) has explored the potential of peripheral regions to overcome these barriers by exploiting various local "soft" factors, such as strong human/social capital, effective governance ("institutional thickness") and balanced business networks. Given the right combination of these conditions, it is argued that even relatively remote regions may develop the characteristics of "innovative milieu" which will allow indigenous micro-businesses to compete successfully despite the centripetal forces described above (Copus, 2004). It is within this perspective that the role of the RDR in supporting the rural economy of remote rural regions needs to be considered.

Apart from the overall paucity of resources allocated to Pillar 2[19], some conclusions may be drawn about current prioritisation of support to diversified micro-businesses in rural and peripheral regions through an examination of the balance of budget allocations to different types of measures. DG Agriculture (2003) has presented a very useful threefold classification of the 22 RDR measures. More than half the RDR budget 2000–2006 was allocated to the group of measures concerned with the environment and land management, a further 38% to a group of measures directed to "restructuring and competitiveness" (i.e. predominantly within agriculture and forestry), and just 10% to measures targeted on the "rural economy/rural community". A review of national plans (Williams, 2004) shows that although the balance between the first and second groups of measures varies somewhat between Member States, the third group consistently receives the smallest allocation. One very straightforward implication of the likely consequences of the MTR and the impact of peripherality is that serious consideration should be given to adjusting the balance of expenditure towards the rural economy/community measures, particularly in remoter regions where extensification and diversification is likely to be the only viable course of action for the majority of farmers. Particular attention should be paid to developing measures to strengthen local business networks (including short supply chains for locally processed quality food products) which have the potential to act as a surrogate for agglomerative advantages (Johansson and Quigley, 2004, p. 165).

In more accessible regions, and in those with more favourable farm structures, the likely trend towards a more market-oriented agriculture will lead to greater need for competitiveness among farm businesses. This would suggest a stronger emphasis on the second group of RDR measures in these regions. However, competitiveness will also in part be influenced by the costs of reaching markets. In this regard, it is important that domestic transport policies of Member States ensure that rural-based food processing industries are facilitated to access international networks.

7.5.2   CAP/RDP and transport policy

Transport and communications infrastructures have a major influence on the spatial distribution of economic activity and also on the underlying dynamics of change throughout the European territory. Project 2.1.1 is concerned with the territorial impacts of EU transport and TEN policies. Here the concern is with the interaction between CAP and RDP on the one hand and EU transport policy on the other hand.

The principal theme of EU transport policy that is of relevance here is the Trans-European Networks initiated in the 1990s. The primary objective of the TENs project is to support the Community objectives of competitiveness and cohesion. Interregional competitiveness is expected to be enhanced through cost reductions resulting from more efficient transport systems. The TENs project

---

[19] The need to "adjust the balance of support" between Pillar 1 and Pillar 2 has already been stressed.

provides new links and improvements to some existing network sections. The project will result in an improvement in both the quantity and quality of infrastructure. By extending the networks into peripheral regions, which are more heavily dependent on agriculture, it is anticipated that there will be greater convergence between core and peripheral regions and, therefore, greater cohesion.

Bröcker *et al.* (2004) have noted that 8 of the 14 priority projects of the TENs programme are located in peripheral regions, but that relatively large improvements in accessibility will translate into only relatively small increases in regional economic activity. In general, it is anticipated that the impact of transport investments on economic development will be larger in regions with less-developed networks than in the regions with dense and better-developed networks. However, they caution that the overall impact of transport investments will depend on the competitiveness of the regional economies: a peripheral area may benefit from better access to markets but its production may also be subject to a higher level of competition from imports. These conclusions are particularly important for agriculture and rural development.

The CAP support system has protected EU producers from more competitively priced imports, and also through the intervention system for some products (e.g. beef) delayed restructuring towards high value added processing and thereby supported relatively inefficient sectors in some regions. The trend towards a more market-oriented agriculture will lead to greater emphasis on competitiveness which will in part be influenced by the costs of reaching markets. In this regard, it is important that domestic transport policies of Member States ensure that rural-based food processing industries are facilitated to access the international networks. Similarly, the marketing of products arising from farm diversification programmes will require transport systems that provide timely and cost effective access to the main centres of population.

The most likely scenario emerging for agricultural production in many regions is one where there will be a relatively small number of intensive and large-scale internationally competitive producers while the majority of farm households will become increasingly dependent on alternative sources of income. The opportunities for additional income will be influenced by trends in the non-farming economy, especially in the services sector and will need to be assisted via comprehensive rural development programmes. This inter-sectoral shift also has implications for spatial patterns of development, with urban centres having a vital role. The nodal position of small towns in respect of several networks is likely to be a critical influence on their level of competitiveness as locations. In order to maximise the opportunities that may emerge from rural development programmes it will be necessary to ensure that the quality of local transport infrastructures is enhanced in order to facilitate spatial integration at the local and regional levels as well as integration with networks providing access to national and international markets. The quantity and quality of the domestically provided local and intra-regional infrastructure is probably at least as important for successful rural development as are the interregional and international networks which are the main focus of EU transport policy to date. Thus, in summary, there is a need for better coordination and closer integration of supranational, national, regional and local transport policies that will enhance the competitiveness of agriculture-based and other rural enterprises.

Chapter 8

8. Ways Forward in Rural Development

8.1. Key dimensions

Rural development as a concept comprises a wide set of notions with different priorities. Given the vulnerable and often less successful economic performance of many rural areas in comparison to urban areas the economic issues and viability is a core question in the light of future development.

It is now common understanding that a pure sectoral approach is not successful in enhancing and stabilising a region's performance; all the same, the notion that rural development goals widely overlap with agricultural policy is still immanent. Though agriculture plays a major role in shaping the rural landscape, it has been made clear that rural citizens, including most farm families, depend on employment and income generated by a complex mix of interacting economic activities. Therefore an integrated, territorial approach is needed to ensure regionally balanced development and social, economic as well as territorial cohesion.

A comprehensive definition of rural development argues that "rural development is increasingly viewed as a territorial concept involving increases in the welfare of rural citizens, including incomes, and quality of life" (Bryden, 1999). This concept in its broad approach can also be subsumed in the concept of "sustainable rural development". It marks a shift from the concept of rural development concerning mainly economic growth and sectoral approaches to a more "holistic" concept covering economic, social, environmental, and spatial dimensions.

The notion that rural and urban areas are distinct territories with distinctive relationships and cultures is challenged by a rising understanding that urban and rural areas are inter-linked in a complex and multidimensional way, a notion which is reflected in the recent discussion of urban-rural relations and partnerships approach as well as the impact of the concept of polycentricity on territorial performance. The various relationships (physical, economical, financial, informal, etc.) are currently investigated particularly by the ESPON project 1.1.2 on urban-rural relations.

8.1.1   **Rural diversity**

The ESDP describes rural areas as "complex economic, natural and cultural locations" (EC, 1999, para 89) which differ markedly from one another in their economic structure and activity, their natural and human resources, the peripherality of their location, their demographic and social conditions, and culture. This diversity is widely perceived as a promising feature in coping with change and in developing new bases for economic and social life. A uniform development trajectory for rural areas does not reflect the actual tendencies under these circumstances. On the contrary, rural diversity is challenged by a number of divergent, place-specific trends and incidents.

Although the forces driving change may be similar across rural (and indeed urban) areas, their relative importance and their consequences will certainly differ from one region to the other. Many rural areas especially in remote and sparsely populated areas appear to be facing particular problems with economic development and adjustment processes. Agriculture as a source of rural employment and income is declining and further job opportunities, especially well-paying jobs, are often scarce. Per-capita incomes are well below national averages and levels of public services and the quality of many kinds of facilities are decreasing.

Working-age people, and especially better-educated younger people, may move elsewhere to find better chances and opportunities, due to the apparent difficulties in meeting their expectation on jobs, educational and leisure facilities. The loss, especially of younger people, along with in-migration of retirees in some places has left many rural areas as "ageing" areas.

This demographic change endangers the rural fabric and leaves, particularly peripheral, rural areas socially and economically neglected, losing the critical mass which is necessary for the establishment and maintenance of facilities, services, and infrastructures. A development that also concerns farm families to a great degree as they rely on a "living countryside" as well, where they can access the services, infrastructure, and supplementary employment they need (Bryden, 1999, p. 8).

On the environmental side a declining population could be a relief of human pressure and an enhancement of natural assets, while on the other hand out-migration and the giving up of mostly small farms in less-favoured regions reduces the variety of historically developed methods of land use and threatens the originality of cultural landscapes. Moreover non-cultivation of formerly utilised agricultural areas may lead to natural hazards such as landslides, avalanches, etc. Physical change of landscapes is also provoked by agricultural intensification processes which tend to lead to more uniform landscapes and a loss of biodiversity which might threaten the highly valued cultural landscape development.

On the other hand, the population in rural areas, particularly in reach of greater cities and agglomerations, is growing steadily, reinforcing the trend towards scattered settlement development and pressures on land use on extended parts of rural areas. Overexploitation, competing demands and interests may threaten the rural diversity as a whole and especially the provision of amenities, cultural heritage features and the environmental performance.

8.1.2 Differential economic performance in rural areas

Understanding the reasons for differential economic performance and more or less competitiveness in rural areas could thus be a key element in devising practical strategies and programmes for sustainable rural development. Tangible factors such as natural and human resources, investment, infrastructure and economic structure have been analysed as the almost unique decisive factors for economic development for a long time. Transport infrastructures, inward investment and towns with additional functions to agglomerations are often mentioned as important conditions for a vibrant rural economy. In the current discussion of recent years less obvious features of an area's economy, which refer to social arrangements and their human participants which cannot be measured and assessed as easily, are gaining importance ("less tangible" or "soft" factors). Recent literature supports the idea that various kinds of social, cultural, institutional, environmental and local knowledge constitute the basic capital for regional development which has high significance when seeking to understand the differences in economic performance. Especially social capital has been addressed as shaping the basic preconditions for successful and lasting regional development. The EU project "Dynamics of rural areas" (DORA) has analysed these dynamics and has identified five areas of less tangible factors:

- market performance (how markets work in practice),
- institutions (how local people perceive and use the institutions which are intended to serve them),
- networks (the personal networks which link public and private sector organisations),
- community (the sense of community as basis for cooperation), and
- considerations of quality of life affecting individual choices (Bryden and Hart, 2001, p. 35).

It is the *relationship* between *tangible and less tangible* resources and how they interact in the local context which gives rise to or conditions different opportunities and constraints for local development. Although mainly tangible resources have long been in the centre of economic development in rural areas recent literature of rural development increasingly refers to the view that "it is not so much the tangible resources themselves that matter for economic performance, but the way the local people are able to exploit those available to them and sometimes to ensure a favourable flow to transfers in their direction" (Bryden and Hart, 2001, p. 45). In other words it is the ability to transform stocks into flows: valorise national and man-made assets, strengthen the economic environment, improve institution capacity. Thus "less tangible" factors determine the efficiency and effectiveness with which tangible resources are used and are most important in making the difference (Pezzini, 2003, p. 5).

8.1.3   The role of entrepreneurs and SMEs

While there are numerous examples of rural places where local (individual and social) enterprises work in a cooperative and flourishing way it is by far more difficult to establish such enterprises in places where they do not exist. As the analysis of the DORA project revealed, there are quite variable conditions underpinning presence or absence of local entrepreneurial capacities. A key issue to emerge in this respect is effective and open governance with a positive attitude to small local enterprises and entrepreneurs and local public institutions with sufficient autonomy to adapt policies and specific measures to assist with the collective needs of local enterprises. Furthermore open and inclusive "soft" networks are positively related to the mobilisation of entrepreneurial capacity and local initiative. Good cooperation between public, private and voluntary actors means that information on a wide range of significant matters circulates widely and freely. Nevertheless, networks have an ambiguous relationship with economic success. In those places where networks appear to be exclusive and associated with a notion of an "elite group" rigid network structures can inhibit or prevent new entrepreneurs.

It is important that social capital conditions, cultural context and the regulation framework induce a positive climate for entrepreneurship, especially in small and medium enterprises "through reducing the costs of information and transactions, reducing risk and uncertainty, creating local public goods which are adapted to support enterprises and encourage new enterprises with positive feedback to the rest of the economy" (Bryden and Hart, 2001, p. 28).

8.1.4   Valuation of rural amenities

It is now commonly agreed that rural areas may contribute to the quality of life because of their particular rich variety of amenities. Thus, when talking about the future of the countryside the cultivation and management of the use of rural amenities is getting to be a key issue.

Rural amenities range from "almost intact nature" such as native forests, desert wilderness or high mountains to amenities which develop from the interaction between nature, and man-made amenities. Most rural areas have been transformed by human activities over long periods of time and these interactions are often important sources of amenities such as farming landscapes or forestry. Man-made rural amenities are expressed furthermore in cultural traditions, historical monuments or artefacts.

In the current discussion of rural development the cultivation of rural amenities is seen as a promising resource because it offers new economic opportunities to rural areas where the economy often lags behind. Rural amenities are strongly associated with specific territorial attributes. Their value stems from the unique features of a given region which cannot be (easily) replaced or exchanged (less-mobile). Thus, it gives the same region a chance to enhance its competitiveness through "cultivating" the place-based social, cultural, and environmental assets.

The increasing value of rural amenities is related to the improved transport links that make recreation in rural regions easily feasible, as well as residential use an attractive prospect, for an increasing part of the population. Above all the valorisation trend has to do with the growing demand on the part of urban dwellers to enjoy amenities in the countryside.

This high valorisation from outside the regions contributes to strengthening the high value on amenities expressed by local people, which symbolises their distinctive cultural identity. The successful cultivation of rural amenities needs both the regional identification of natural and cultural amenities, with the desire to preserve it, and favourable structures of decision-making processes in the region. This includes for example local institutions with capacity to recognise the market value and who are able to organise and coordinate supply and promotion of the specific local amenity.

However, the relationships between amenities and rural economic development are not as simple and straightforward as they seem to be and the outcomes vary considerably. The OECD describes three types of relationships between amenities and rural development (OECD, 1999):

- *Synergy* is apparent when preserving amenities supports development. For example, sustainable tourism stimulates economic activity and the arrival of new populations in the countryside without destroying the underlying amenity.
- *Antagonism*: i.e. when preserving amenities dampens the rural economy. Sometimes, preservation reduces rather than increases human activity so that economic growth is constrained. But on the other hand an excessive use, caused e.g. by mass tourism, endangers amenities provision and the dependent economy development.
- Interdependence is apparent when economic stagnation negatively affects amenities. Man-made amenities require at least some economic development. When a rural area is depopulated because of economic decline, the associated cultural values and environmental resources are both threatened.

Clearly, the aim of exploiting amenities for rural development is to enable and support synergy effects. Therefore a primary objective of amenity policy should be to "first establish supply at a level that matches existing demand and assumes demand by future generations and, second, protect amenities from irreversible damage so that a range of future uses is ensured" (OECD, 1999, p. 33).

8.1.5   Institutional framework for territorial policies (rural governance)

Rurality is no longer synonymous with agriculture nor is it synonymous with decline. Given the varied nature and the multiplicity of challenges and developments that rural areas are facing, they cannot be addressed solely by agricultural policies. A key policy issue is, therefore, how to get greater positive inter-relationships and synergies between sectoral policies in different spheres like economy, environment and social affairs. A multi-sectoral approach which includes the territorial dimension is thus crucial to respond to rural demands.

In recent years the appropriate level of decision-making with regard to rural development became a matter of discussion. The diversity among rural places and their characteristics makes it very difficult and inappropriate to design policies at a central level (EU or national) which can take into account locally specific needs as well as geographically balanced economic development for a nation (and for the EU). To enable the use of knowledge dispersed in the various rural areas, to secure the necessary consensus for policy implementation and to strengthen effective participation in decision making the implementation of an active role for different levels of governments (local, regional, national, supra-national) seems to be necessary. Thus, countries increasingly pursue decentralisation and devolution of policy making towards regions and localities in order to better meet diverse needs and conditions found in rural areas and tap local knowledge and other resources. The demands on rural policies are that they should both comply with bottom-up principles and be appropriately linked to top-down structures of support. The EU-LEADER programme for example, among other innovative local pilot programmes, seems to be a successful approach to stimulating such integrated development using the endogenous potential of the area.

In relation to institutional needs it has to be taken into account that the development of rural areas is based more and more on interactions between different adjacent areas and interregional exchange across greater distances. Cooperation between communities and the putting in place of horizontal partnerships between public and private actors reflect a place-based approach to development and cut across traditional administrative boundaries. Within these new forms of territorial development, municipalities constitute micro-regions, territorial pacts, and different sorts of consortia that tend to become the interface through which policies will be implemented. These most flexible forms of governance permit governments to better exploit local complementarities and to ensure continuity in infrastructural development through sharing of public investments.

In this context, it is important to analyse the exact role of administrative units (municipalities, districts, regions, etc.) and to propose a framework for maximising their contribution to rural development. Pezzini (2003, p. 12) focuses on the following principal issues which should illuminate the interplay of different levels of policies and sectors as well as the process of involvement of local people:

- Structures and rules that governments put in place in order to promote or support local initiatives.
- Inter-sectoral coordination and coherence.
- Finance and incentives (own tax revenues, contributions, subsidies, etc.).
- Contracts and the process of negotiation.
- Learning processes to strengthen their role in the design and implementation of development policy, which raises the issue of the human resources available and capacity building.
- Actors involved (including civil society participation and role).

From the perspective of the national governments, rural policy design and implementation in the recent past has to do with the fact that local and regional governments have been brought more strongly into the picture. This recognition of the crucial role of local-level decisions is shared both by local and national governments. However, also at the level of central government, there is an intensified need for acknowledgement of rural issues in shaping national territorial strategies and very often there remains considerable room for improvement in coordination of various ministries and departments responsible for policies affecting rural development. Some key elements of these adjustments and policy reform strands seem to be (Bryden, 1999; Pezzini, 2003): policy-proofing by a senior inter-departmental or inter-ministerial group to ensure that all policies (for example in addition to classical regional support schemes, and agriculture and rural development policies, policies for housing, transport, telecommunications, health, etc.) consider the rural dimension; the allocation of rural coordination responsibilities to one senior ministry or department which must chair the inter-departmental or inter-ministerial group; and the establishment of national or supra-national networks of local partnerships (as for example in the European LEADER Observatory) which exchange information, run training seminars, and provide documentation on "good practice".

8.1.6   Determinants of good practice

The assessment of policy application requires a framework to relate experiences from case studies. In general, the type of place-specific action makes it rather difficult to compare implementation impediments, and factors of success or failure. Nevertheless, a series of principles and general considerations was established prior to case study execution to guide these. The following items highlight the approach and key elements of an integrated concept to rural development. According to the nature of the instruments, some elements are more appropriate for specific measures than others.

- "Good Practice" includes "Good Structures", i.e. the shape, size, etc. of the organisations and agencies that promote rural development. Together, these might be termed "good institutions", where "good" means "appropriate, given policy goals and available resources", and (in accordance with social science generally) "institutions" means both organisations (e.g. agencies, firms) and rules (e.g. the state of contract law, corruption, behavioural norms, etc.).

- "Good Practice/Structures" might well include organisations in the private (and voluntary) sector, i.e. entrepreneur groups, and businesses. However, the prime focus of the project is on CAP impact and thus we did not concentrate on non-organised individuals (e.g. consumers, maybe also individual entrepreneurs) in our investigations of "good practice".

- Depending on the nature of the instrument assessed and the information available within each case study, it may be possible that only some of the following items are addressed (or parts of them): eligibility for policy support, i.e. private "structure" should match policy measure; willingness and ability (e.g. education, time/distance) to participate in policy administration (public-private partnerships, representation on policy committees); willingness and ability (e.g. financial) to take up policy measures; appropriate (i.e. "good") business ownership and structure, e.g. (possibly) local ownership rather than distant or (inter)national ownership; single rather than multiple sites/plants; competition rather than quasi-monopoly; supportive structures for policy-induced diversification products; appropriate organisations to reflect increased valuation and demand for preserved and improved environmental conditions, e.g. cultural landscapes.

- "Good Practice/Structures" in rural development policy can be defined more or less independently from the goals of any particular rural development measure. The elements of good practice/structures are experienced across rural regions and case studies are oriented towards revealing interesting cases of good practice in the respective fields.

- The known/assumed elements of good practice/structures include: adequate eligibility (enough potential take-up); efficient administration (public) and business management (private); adequate consultation (before policy implementation is finalised); adequate advice (while policy is implemented); adequate support structures; appropriate co-funding requirements.

8.2. A selection of good practices in rural development

The application of rural development programmes throughout Europe has been described as very diverse. The aim of this chapter is to highlight some examples that present either innovative approaches or representative use of the framework provided by RDR. As Dwyer *et al.* (2002) have found, innovation in RDP across Europe has occurred at the level of both programme design and resourcing, as well as at the level of individual project and initiatives. Many examples would demonstrate notable flexibility and tailoring of measures to meet local circumstances and potential. In this respect it is notable that the strongest examples relate to developments which precede Agenda 2000, either coming from previous experience with Structural Funds or from essentially separate national or more local initiatives. This finding has been underlined by various research on local initiatives and relevant factors of success, which particularly pointed to the social capital base (Arnason *et al.*, 2004) and the institutional development within the local area (Dax and Hovorka, 2002). Thus the examples of innovative approaches cannot be detached from support mechanisms and structures, at all administrative levels, impacting on the actual performance of rural development.

Examples have been selected so as to address the different instruments analysed and reflect the on-going policy discussion on future reform. As they are examples, of course there would be a lot of additional, more detailed information. Analysis of "Good Practice" includes analysis of the institutional context and behaviour for Pillar 2 and the selected case studies highlight particularly programmes and initiatives with a more decentralised organisation trying to support endogenous development of rural areas and with high participation in less-favoured areas. The cases presented below include:

1. Differentiation of compensatory allowances for LFA in Austria
2. Ireland Rural Environment Protection Scheme (REPS)
3. CTE (contrat territorial d'exploitation), France
4. "Cheese Route Bregenzerwald", LEADER, Austria
5. Rural Tourism in Italy
6. PRODER Andalucia, Spain
7. POMO and POMO+, Finland

8.2.1 Differentiation of compensatory allowances, Austria

The landscape of Austria is characterised by the high proportion of mountain areas (70%). Together with other less-favoured areas and areas of specific handicaps almost 80% of the total land area, and 70% of the utilised agricultural area (UAA) are classified as LFA. This situation has already led, in the 1970s, to the first support measures for mountain farming, and in the same period to the start of comprehensive mountain regional programmes. From the beginning, linkage of the different policy strands has been deemed to be crucial in mountain development, as mountain farming was regarded as having a particular role in safeguarding the sensitive ecosystem and thereby the multifunctional landscape and working and living space.

Although Austria has a particularly big rural development programme (with 9.5% of EU funds for RDP spent for the Austrian programme), the budget for the measures relating to LFAs reaches approximately 29% of total Austrian RDP costs. This underpins the long-standing commitment for this type of support in Austria and contributes, together with the agri-environmental programme (ÖPUL), to meeting the objectives of compensation for special services provided by LFA farmers, preservation of assets, and improving competitiveness of agriculture in these large regions (Hovorka, 2003).

The gradual extension of the compensatory allowances scheme to farms outside the mountain area increased the interest in the differentiation of compensatory payments. In general, it has been implemented through a base category and higher support levels for mountain farmers. The base category which is relevant for farmers in other less-favoured areas, in areas with specific handicaps and those farmers in mountain areas with limited production difficulties (particularly on flat areas in the valleys) is set at about 40% of the maximum level of support. Compensatory allowances for mountain farmers are differentiated through a complex system of measuring the agricultural production difficulties and regional situation of affected mountain farmers.

Since the early 1970s a differentiated classification system (of 4 groups) has been the core element for defining the support payments for mountain farmers. The main criteria for the classification were the climatic conditions and the "internal transport situation", i.e. the proportion of agricultural area of the holding that had a gradient of at least 25%, or of at least 50% for the farms with highest difficulties (group 4). This differentiation of mountain farms operated until 2001.

The change to a more differentiated payment structure was planned and prepared during the 1990s and a revised classification system (Tamme *et al.*, 2002) has been applied since 2001. In addition to the most relevant aspect of "internal production handicaps" (already used in the former classification system), further indicators of external transport situation (accessibility of farm, regional situation with regard to public transport and general service, factors of isolation and overall regional economic performance) and soil and climate indicators are included. The application of the new system has been made possible through data gathering on farm plot specific information for all mountain areas. Moreover this information is updated and changes in farm management are included in the yearly calculation of compensatory payments. By summing up the relevance of 15 separate indicators a composite degree of production difficulty for mountain farms is achieved. This can have values between 0 and 570 points and thus provides a classification of individual farms which avoids adverse effects at the limits of groups.

The transparent system makes it possible for each individual farmer to check his classification and the particular profile of production difficulty. With the application of the system the long work on the establishing of the system should pay off, as the yearly up-date and the more accurate system receive greater acceptance among farmers than previous crude classification of groups.

8.2.2   Rural Environment Protection Scheme (REPS), Ireland

Context
Post-1973 modernisation of agriculture (Ireland's accession to EEC) – economic and social benefits but negative environmental impacts:

- pollution from silage effluent and animal slurry
- excessive fertiliser applications- eutrophication of lakes and rivers
- contamination of ground water

- land reclamation and drainage- destruction of wildlife habitats
- loss of sites of historical and scientific interest
- visual intrusion of farm buildings on landscape
- increased livestock numbers; increased levels of methane gas; breaching Kyoto targets.

Very little agri-environment support prior to 1992
REPS devised by Department of Agriculture and Food and launched in June 1994

Objectives

- To establish farming practices and production methods which reflect increasing concern for conservation, landscape protection and wider environmental problems/issues
- To protect wildlife habitats and endangered species of flora and fauna
- To produce quality food in an extensive and environmentally friendly manner.

Eligible farmers entitled to payment of €151/ha to max of €6040 (1994)

Application

- 11 horizontal measures
- 6 supplementary zonal measures.

Characteristics: universal availability, voluntary, restriction of payments to <40 ha, inclusion of training element

Adoption of REPS

Initial target approx 45,000 farmers (25% of total), 1.3 million ha
2003 (Oct) 37,000 participants (29% of total farmers); 1,312,200 ha
Highest participation rates principally in areas with small farms and extensive farming systems; low rates in most intensive farming (and most damaging) areas

Conclusions

- REPS of greatest benefit to low-income small farms in more marginal farming regions
- Compensation payments not sufficient to encourage intensive farmers to adopt
- Evidence of improvements in farming practices leading to reduced environmental impact (reduced application of inorganic nitrates and phosphates); very little evidence of environmental enhancement (especially in relation to habitats and biodiversity)
- There are particular concerns about lack of monitoring and the absence of specified targets.

8.2.3   **CTE (contrats territorial d'exploitation), France**

The most prominent innovation in the French RDP was the CTE. They were developed from the French agri-environment measures introduced after 1992. It comprised a combination of national measures, available over the entire national territory, and "local operations", drawn up by the regions in response to local circumstances. These local operations which eventually numbered over 300, were largely considered a success both in terms of their acceptance by the farming community and in terms of their environmental and agronomic impact. The innovation of the CTEs was to group these local operations into a series of generic component "measures", and to link them into a wider multifunctional and rural development framework. The result was a national catalogue of around 80 generic agri-environment measures and over 150 generic contract types, tailored to particular agricultural or environmental circumstances. Appropriate measures from this catalogue are then combined with other measures available for farm development (notably farm investments, setting up young farmers, and early retirement measures, to produce generic contracts at the Département (county) level, comprising both voluntary and obligatory components.

The CTEs have the following characteristics:

- they are contractual, between the farmer and the state
- they explicitly link farm development to environmental management
- they address the multifunctional role of farming in that they offer a package of measures designed to respond to the economic, environmental, territorial and social role of farming
- they are based on a "bottom-up" approach in that individual contracts are constructed to fit local circumstances
- they encourage collective local action – CTE can be drawn up and piloted by groups of farmers in response to specific local demands and concerns.

Inherent in the CTE approach is the decentralised implementation. The counties' prefects determine the measures to be employed in the CTE available in the county. Local Chambers of Agriculture have a key role in establishing local contracts and encouraging farmers to apply for collective CTE projects. Finally a regional commission must approve both the model CTE measures available and the individual CTEs within the county. These new bodies include local political leaders, farmers and representatives of the local agri-food sector, territorial development agencies and local environmental organisations.

Too high expectations and restrictive evaluation led in 2003 to the transformation of the scheme into CAD, contrat d'agriculture durable. However, CTEs signed until July 2003 have amounted to almost 50,000 farm holdings, representing 12% of French professional farmers (Vindel, 2004). The rate of contracting is reported to be higher in the east and south of France, and even higher in less-favoured areas, which is assessed as meeting one of its objectives.

With all its difficulties in implementation, due to the administrative and technical complexity, it has favoured environmentally friendly practices and raised the awareness of the environmental issues among farmers and their organisations.

8.2.4   LEADER "Cheese Route Bregenzerwald", Austria

The "Cheese Route Bregenzerwald" was the strategic lead project of the LEADER II programme in the most western province of Austria, Vorarlberg. The main objective was to emphasise the uniqueness of the region's products (especially cheese) and to increase the region's value added of cheese production by about one-third (from €4.3 million to €5.8 million), thereby contributing to assuring the livelihood of the rural population, reducing the quantity of commuters and helping to create new jobs in tourism and trade. 191 members from the fields of agriculture tourism, alpine dairies, alpine pasture management, accommodation services, trade and commerce formed the largest sales consortium in a rural region in Europe. The association established an organisation where all the ideas to be implemented in the project were brought together, providing a service point for its members. Responsibility for the activities was with the "Regional Development Bregenzerwald plc".

Right from the beginning (and even before the "project" started) the importance of cooperation has been emphasised by representatives of different sectors, particularly from the agriculture and tourism sector, and therefore the preconditions for a multisectoral cooperation can be assessed as substantially favourable. Specific requirements and needs of the region have been discussed in numerous meetings of (key) actors and also in lectures with inhabitants of the region, leading to increased consideration of these aspects in the concept.

The strong personal involvement and inter-linkages of the diverse organisations participating in the process contributed to the high degree of stability and willingness to cooperate. Thus, although the public sector held the overall responsibility and initiative for the projects, the private sector came up with essential ideas and, as they work for their own livelihood, also financed a significant share of the implementation costs. There was also a strong commitment between the members of the cooperation towards further developments of the "Cheese Routes". The pilot project induced the realisation of a remarkable number of related projects, which positively influenced and supported each other, like collective investments in facilities for the preparation and presentation of cheese and in innovative products (e.g. the projects "Käsezwickel", "Käseträger", "Käse&Design"). The association "cheese route" is still in charge of public relations and concerned with improving product presentation, e.g. via its web page. Members are primarily dairy farmers, Alpine dairies, restaurants, retail shops and stores, but extend to other actors as well.

Main factors contributing to the performance achieved through the measure were the holistic concept, inclusion of multiple beneficiaries, participation as a guiding principal for regional governance which is based on a long tradition of citizens' action, an innovative multi-stakeholder partnership, as well as the integrated marketing concept which was able to establish a new high quality brand enhancing positive development for regional and supra-regional sale of products.

The consequences of the project are highly favourable. The regional economic performance could be improved as well as the livelihood of the participants. The long tradition of cooperation could even be strengthened through the large number of participants and different professional backgrounds in the project. Traditional agricultural exploitation of the landscape, which is not only without risks but a precondition for the cultural landscape quality, could be widely maintained. The region Bregenzerwald is currently applying for admission as a UNESCO world heritage region (within the LEADER+ programme).

8.2.5   Rural tourism, Italy

Agri-tourism has long been widespread in some European regions. With increasing diversification a rising number of farm households have engaged in tourist activities and gradually expanded the offer, reflecting the shifts in demand trends. Particularly the efforts of local development programmes, like LEADER, have enhanced the activities and brought about a variety of new activities. Italy is one of the countries where both a highly intensive mass tourism and agro-tourism activities can be found. A manual on agro-tourism within LEADER II reveals the wide range of regional regulations and activities possible under the scheme (Hausmann, 1996).

Agriturismo (farm tourism) offers significant advantages for the province from a sustainability point of view. It broadens the types of tourist attractions and activities offered by the province. It appeals to tourists eager to learn more about local cultural patterns and economic activities, providing a stimulus for forestry and environmentally friendly activities. And because the benefits are more evenly shared throughout the province, it plays a revitalising role in the most deprived areas, generating additional income for farm household and local communities with few other substantial economic activities. A territorial review of the province Siena carried out by OECD has underpinned the strategic role of agri-tourism in this highly advanced tourist region (OECD, 2002c). Based on a series of qualitative trends in tourism more sustainable types of tourism are looked for. Agri-tourism has developed over the last decade as a particular highly demanded type of tourism with constantly high increase rates. The province of Siena is the second most important area in Italy of *agriturismo*. Bolzano, Siena, Perugia, Florence and Grosetto are, in decreasing order, those with the highest concentration of *agriturismo* units, together accounting for 41% of the national total (ISTAT, 1998). In Siena the increase led to a situation where *agriturismo* already offered 32% of the areas tourist beds (in 1998). In nearby province of Perugia a similar positive trend for agri-tourism can be seen (Ventura *et al.*, 2001).

The vitality of agri-tourism can be explained by several factors. On the offer side, the need to diversify agricultural activities and the direct and indirect incentives deriving from the Rural Development Plans and LEADER programmes have played a significant role. The opportunities offered by Tuscany's regional *agriturismo* law have also strongly supported farm accommodation, giving specific status to farms where tourism activities do not account for more than half of the invoicing (similar farm-specific rules are relevant in other countries, like Austria). This entitles them to a preferential tax treatment, a 4% rate compared with an average rate of 27% for other types of accommodation. It is important to underline that *agritourismo* provides an additional income, both through room and board sales and through direct-to-consumer sales of agro-food products (cheese, wine, olive oil, fruit products, vegetables, meat and poultry). Increasingly organic farms are involved in agri-tourism activities. All over Italy, 63% of agri-tourist units offer some kind of gastronomic service which explains the particular attraction of this type of tourism.

On the demand side, the growing popularity of countryside tourism has inspired the farm operators to engage in these activities. A great part of Siena's landscape is agricultural, highly aesthetic, with a variety of hills, plains and woods, and many ancient farmhouses. It is, however, important to mention that in order to provide an adequate range of tourism services in a region, coordination and networking must be integrated into local and regional promotion networks. This includes also use of internet information and reservation facilities.

Another case of agri-tourism development in Calabria shows the situation of a region where this type of tourism has only achieved minor importance up to now (Privitera, 2004). A sample carried out within farm households having applied for Community aid with regard to agro-tourist activities reveal similar trends as in other regions, and a wide range of services offered by these farm households. There are, however, considerable weaknesses in relation to making use of the good development potential. It is the regional strategy to improve the regional tourist offer by increasing the variety and quality of different types of tourism, including agri-tourism.

8.2.6   PRODER: Andalucia, Spain

The PRODER scheme provides a good example of where a Member State has decided to use an established local delivery mechanism to implement various RDP measures in a flexible way, utilizing a bottom-up approach. The programme consists of a set of implementing measures for endogenous development of rural areas and is the main case of mainstreaming the LEADER method up to now.

PRODER (Programa Operativo de Desarollo y Diversificación Económica de Zonas Rurales) is the Operational Programme for the Development and Diversification of Rural Areas and was introduced as part of the 1994–99 programming of Structural Funds for Objective 1 regions in Spain, as a horizontal programme run by the central Ministry of Agriculture. The idea was to replicate the LEADER approach, and extend it to more areas. For the 2000–2007 period, the programme was extended to Non Objective 1 regions, where each region has included a PRODER-type scheme in its regional programmes, sometimes under joint management with the Ministry of Agriculture. Many of the PRODER areas overlap with LEADER regions, but not all. The number of local development programmes increased from 101 PRODER1 programmes (1996–99) to 162 PRODER2 programmes (2000–2007) in 12 regions of Spain. Overall public funding for the current PRODER programmes amounts to €828m, of which the greatest share is allocated for Andalucia (€212m).

In Andalucia (Objective 1), the equivalent scheme is called the Global Grant for Endogenous Development of Rural Areas. It takes the form of a global grant under Article 27 of the Structural Funds Reg. 1260/99. The scheme operates throughout Andalucia, although the budget is allocated preferentially to districts with significant economic, environmental and equal-opportunity problems (Dwyer et al., 2002). Funding for the local action groups comes from both EAGGF Guidance and ERDF. Actions eligible for funding from ERDF are:

- operational costs of the local groups
- support services for crafts, commerce and hostelry
- small industrial estates and fairs
- improvement of towns and village centres
- construction of cultural centres
- support for small businesses
- studies and advice on rural development.

Actions eligible for funding from EAGGF are:

- actions related to farm diversification and/or in predominantly agricultural territories
- basic services and information provision to the population
- cooperation projects
- conservation of cultural heritage (buildings, villages)
- productive activities compatible with environmental conservation and/or aiming at the protection, restoration and valorisation of natural resources and landscapes
- development and improvement of infrastructure related to farm production (e.g. farm roads, livestock units, electricity supply)
- promotion, improvement and diversification of the rural economy, in the agricultural, craft and tourist sectors.

The importance given to PRODER in regional programmes varies considerably. Thus Castilla y León, Andalucia and Asturias dedicate relatively significant budgets to the scheme while others, like Castilla-La Mancha, allocate a relatively modest amount and exclude agricultural actions from the support.

In the case of Andalucia it is important to point out the special interest from the regional authorities to develop PRODER as a similar approach to the LEADER method. The learning process has been more successful than elsewhere, in the sense that PRODER2 is conceived in a different way than in the rest of the country, as a complement to LEADER+ measures. Andalucia is also applying another important innovation, the "homologated groups", which are LAGs officially established to undertake the implementation of a set of policies in their areas. "This is a clear example of the success of the territorial approach, which means more participation of the local actors in the design and implementation of different policies in rural areas. At this moment, PRODER2 and LEADER+ in Andalucia, with their 'homologated' LAGS, are an interesting experiment which have to be followed in the near future, as a real projection of the LEADER method" (Esparcia and Noguera, 2004).

8.2.7   POMO and POMO+, Finland

The Finnish abbreviation 'POMO' stands for Rural Programme Based on Local Initiative. The first phase of POMO (1997–99) was introduced as the national extension to the LEADER II programme. Like LEADER II evolved into LEADER+ for the new programming period (2000–2006), POMO became POMO+ respectively, although it was not originally envisaged that there would be a national follow-up.

Both POMO and POMO+ would have an integrated, multi-sectoral nature to finance a wide range of measures and activities as per the development plans of the respective LAGs. However, POMO did not allow funding for individual business enterprises while there is, in principle, no such limitation in POMO+. But most LAGs operate mainly through collective projects and indirect business development by building capacities and improving the operating environment. While there were differences in the implementation model between POMO and LEADER II, the new programmes are managed along the same lines: the Ministry of Agriculture and Forestry as the managing authority and the Rural Departments of the Employment and Economic Development Centres as paying authorities in relation to individual projects.

The evolution of POMO is viewed as a particular milestone within the LEADER mainstreaming process in Finland. With the introduction of the programme in 1997 an additional 26 LAGs were selected which corresponded to the effective demand for a wider intervention than LEADER II. This brought the number of LAGs in 1997–99 up to 48. The second milestone was mainstreaming of LAGs into the Structural Funds and launch of POMO+ as a complementary national programme to LEADER+. POMO+ allowed funding for 7 LAGs bringing the total number of LAGs in the period 2000–2006 up to 58 which cover practically all rural Finland. The differences between the two programmes (e.g. lower funding level in POMO) have been harmonised along the lines of LEADER+ and there appears to be a fair amount of synergies between the programmes.

POMO has achieved its objectives to create or safeguard 800–850 jobs, contributed to the creation of 200–250 micro-enterprises, and proved particularly effective in diversification (mainly rural tourism). The established network of committed actors at all levels of rural development work have constituted a major internal success factor for mainstreaming. It was primarily a question of communicative interaction between administration and the civil society to carry through the bottom-up approach of the LEADER method.

Although the success is highly related to context-specific operationalisation two critical areas are highlighted for the Finnish case: first, the most challenging aspect of the LEADER method is that LAGs remain autonomous and empowered; second, despite well-started networking there seems to be a lot of unrealised potential in transfer of good practices and innovation in both national and transnational contexts. Hence, more attention should be drawn directly to upgrading learning processes which could be considered as one of the key features of the LEADER method.

8.3. Mainstreaming the LEADER approach

The LEADER Community Initiative is one of the four remaining EU CI for the period 2000–2006, but it has a very limited budget (€2.02 million), compared to overall Structural Funds and CAP. Nevertheless it is the programme which is particularly related to the concept of integrated rural development, and provides a multitude of good/bad examples of rural development under different contexts. Moreover this has a crucial impact on the political discourse and effects also on the discussion of regional development of peripheral areas. As such regions are very content to dispose of an instrument with a highly experimental character where innovative approaches could be started, too. In particular:

- Beyond the economic sphere the programme is important for other spheres of rural life and policy, as the regional strategy development touches upon a much wider field of sectors than just the prime sectors usually addressed by Rural Development programmes.
- The development of a regional strategy is an important aim and achievement in itself. This can be used by the regions for further activities and spreads out to other sectors' activities (see for example case study Spain).
- It provides a flexible programme structure which has to be adapted to the context of the rural regions, and has achieved interesting results for small-scale regional development. Numerous case study descriptions (beyond those carried out under the ESPON programme) elaborate on the starting period, the difficulties and outcome of initiatives. Some of them also underline the requirements for the successful application and institutional prerequisites, including the following characteristics of action-centred networks: flat, flexible organisational structures involving teamwork and partnership; equality of relationships among relevant stakeholders; vision and value-driven leadership, and emphasis on participation and organisational learning.
- The core of the programme is the emphasis on the multi-sector approach which requires a high commitment by participants to overcome institutional

and deeply entrenched personal difficulties with regard to cooperative activities and new ways of organisation at the local level. This discussion has turned out to be very important for the discussion of "regional governances".

- During the LEADER period – as it is analysed in the Case of Northern Ireland – there is evidence of a increasing level of rural development know-how and an improved capacity of partnerships to deliver programmes for rural development. LAGs no longer see themselves mainly as provider of local funding on a project-to-project basis, which often resulted in a "scatter-gun" approach to development. This change to a programme-driven approach allowed LAGs to manage and target resources in a more effective manner rather than simply react to various project ideas (Scott, 2004).

- The inclusion of social, cultural and environmental concerns is regarded nowadays as part of good practice in regional policy; part of it can be attributed to LEADER experiences and the concern for harnessing natural and cultural assets in rural areas as a prime development potential for many rural areas.

- Participation is not satisfying everywhere. In particular, different groups of society are underrepresented and strategies are the expression of the discussion process and power relations of decisive stakeholders in the areas. An enlargement of the groups addressed and integrated in the process is one of the actual objectives of the current period (e.g. stronger participation of women, young people etc.) and would also increase the effectiveness of the approach.

The future integration of LEADER+ into the rural development programming (*mainstreaming*) as outlined in the Third Cohesion Report might have severe implications on the administration and contents of LEADER activities. The specific features of the Community Initiative should be maintained (and elaborated) in order to build on its potential. LEADER is very effective in creating new links between local actors and stakeholders (re)building trust across contemporary social divides and sectoral points of view. However, this cooperation and the development of common strategic planning needs time and LEADER issues like multi-sectoral integration, networking and transnational cooperation between rural areas were often too ambitious for the LAGs (transnation cooperation) or were achieved only by the more advanced groups. For example, the successful implementation of multi-sectoral integration was an effect of certain preconditions and external influences, e.g. a favourable administrative context; a thriving and diversified local economy; a viable, dynamic, representative mixed partnership and a strong strategic orientation in the local action plan, rather than of LEADER activities (ÖIR, 2003, p. 26).

Within the mainstream programming, there should be an opportunity for (newly) defined regions to get together, recall their endogenous potentials and explore new ways of development according to the respective situation in the rural area. In particular, new-founded LAGs in the new Member States will need space and time for experimenting with authentic ways of development. A

LEADER-type mainstream programme will require considerable resources for capacity building, negotiation and organisational development, and thus a period of reduced economic cost-effectiveness is to be taken into account. An increase in efficiency in programme implementation and especially in disbursement of funds is to be expected in later stages (ÖIR, 2004).

On the other hand, more experienced LAGs should be supported to maintain and improve their development structures. The focus could be to support their efforts in the direction of multi-sectoral integration, networking and transnational cooperation between rural areas, all features which need already existing and functioning internal networks.

ÖIR (2004) aimed exactly at analysing the issues of if and how (far) individual LEADER features or the method as a whole are applied in the "classical" rural development measures. The overall conclusion was that the LEADER method is applicable to the whole range of rural development measures, although incidence of LEADER-type measures varies considerably between RD measures and Member States. There are positive outcomes in respect to (i) regional added value (including for example the development of soft factors like participation of different groups of actors or an efficient decentralized management and financing, though this depends on the concerted interplay of authorities and institutions at various levels), (ii) production of synergies with other regional development measures particularly in Objective 1 areas, (iii) and feasibility of the different features of the LEADER method.

Among the 22 RDR measures, the most frequent mainstreaming has taken place within the scope of Article 33 on adaptation and development of rural areas. But, in general, strong mainstreaming includes farm investments (Art. 4 to 7), setting up of young farmers (Art. 8), and investments in marketing and processing (Art. 25 to 27). Less frequent, but successful mainstreaming is recorded from agri-environmental measures (Art. 22 to 24) and forestry (Art. 29 to 32).

There is also quite a considerable difference in degree of intensity of mainstreaming within the Member States as well. Strong and full mainstreaming appears on the one hand as a pan-territorial approach, by which the administration seeks to offer a LEADER-type programme to all rural areas and actors (ES, IE) and/or on the other hand as a process of structural transformation (FI, IT). The case studies of the PRODER programme in Spain and the POMO programme in Finland refer to the strong (resp. full) mainstreaming of LEADER. Weak and light mainstreaming can be observed, either where the LEADER approach is used for niche programmes with specific measures (e.g. for remote areas, for rural tourism) or where the method is infiltrating on a broad basis, but slowly and incrementally (e.g. DE, DK, FR, SE).

Important difficulties in mainstreaming the LEADER approach (which should be considered in the future concept of mainstreaming LEADER) arise particularly with regard to following areas:

- Problems related to programming rules and regulation, e.g. the EAGGF-Guarantee implementation rules limit the eligibility of non-agricultural

activities; the annuality principle is not appropriate for project-oriented funding.
- Political and institutional hindrance in Member States where in some cases decentralised management and financing through local groups is not backed up.
- Administrative barriers related to routines of a sectoral perspective and of large-scale payment operations
- Problems related to local social capital as local actors need time to build up the strategic and operational capacities necessary to design and implement local development strategies.

The report concludes that there is a strong case for mainstreaming the LEADER method or some of its features into all rural development measures. But, especially with regard to a favourable governance context, certain amendments should be considered and the introduction of three types of interventions is recommended (ÖIR, 2004, p. 91):

- in programme design: removing the administrative, structural and capacity related barriers to mainstreaming the LEADER method at Community level and to take into account the vast differences between rural areas in the EU-25;
- in programme implementation: offering strong incentives for mainstreaming the LEADER method in national rural development programmes;
- in programme support: setting up a European networking device (to support transnational networking).

It seems that the discussions at the Salzburg Conference on Rural Development in November 2003 and the recommendations on mainstreaming the LEADER method were taken into account, at least partially, in the proposal of the Commission for the future rural development policy, adopted on 14 July 2004 (EC, 2004c). It includes the activities previously financed under the LEADER initiative in the new rural development policy as a fourth implementation axis. The requirement to set up a LEADER element for the implementation of local development strategies of local action groups for each programme implies a particular concentration on these measures and increases the attention for spatial considerations of the rural development programmes. As a minimum of 7% of programme funding is reserved for the LEADER axis the current level of LEADER activities may even be extended and will be applied to all rural regions across the EU-25. The proposed framework provides the basis that the LEADER model can be applied on a wider scale by those Member States wishing to do so and can be used for region-specific strategies, while for the EU as a whole continuation and consolidation of the LEADER approach will be safeguarded. This might allow Member States to address more directly spatial aspects of future rural development policy.

8.4. Summary

There is marked trade-off between the main objectives of CAP and spatial policies. However, this relationship is not stable and has changed substantially over the recent decade. A more thorough inclusion of spatial aspects in considerations for agricultural policies has led to the extension towards a Rural Development Policy encompassing both the preservation of agricultural land use through a competitive farm sector and the sustainable development of rural regions, which includes the whole rural economy far beyond the agricultural sector. The inter-linkage between the different sector activities, actors and institutions has been increasingly addressed in some of the innovative applications of rural development schemes. This chapter on good practice therefore tried to highlight some of the cases and experiences instructive for the future development of a balanced strategy for CAP and rural development.

It is widely accepted that this process requires a multi-dimensional approach, which reflects the particular regional (local) contextual situations. The instrument which has been discussed under these circumstances most intensively is the LEADER Community Initiative. Despite its limitations, it was therefore also argued to mainstream the LEADER approach into the rural development policy and adopt its basic principles as guiding elements for future rural development measures. Much of the criticism on the limited impact of rural development measures is about its marginal role in financial terms within the overall CAP (Van der Ploeg, 2003). Mainstreaming of LEADER and further shifts of the EU framework of rural development measures, as proposed by the Commission in July 2004, might address more directly and effectively the diversification potential of farmers and rural regions in Europe.

It is, however, important to extend the spatial aspects to all CAP measures in order to achieve an analysis of the spatial relevance of the components and the whole package of measures. Examples presented here refer to the most widely used and relevant measures of Pillar 2: the less-favoured areas scheme and the agri-environmental measures. Moreover, horizontal and vertical integration of measures in an innovative way includes to a high degree the spatial aspects of farm management and related activities. The particular case of diversification is presented through the case of rural tourism which is by far the most popular non-agricultural activity taken up by farm households.

Good practice experience confirms the findings from earlier scoping studies on the future of rural development, the need for policy changes and the role of institutions and mobilising local actors in this process (Baldock *et al.*, 2001; Dwyer *et al.*, 2002; Bennett, 2003; Buller, 2003; Saraceno, 2004). These can be summarised with following the items:

- The analysis of different cases reveals that application of RDR comprises a wide set of different strategies. These partly reflect the specific regional contexts, partly are influenced by historic experience and national priorities and understanding of rural development. In this regard they make use of the innovative element of the RDR to varying degrees, and although there are some good examples, the potential is generally not yet being fully realised.

- All measures are applied according to the regional context. This spatial feature is to be included in the formulation and implementation of the measures. From this it results that differentiation of measures which reflect the regional situation and/or the farm type and production difficulty are important for the acceptance and the effectiveness of programmes.

- Effective rural development policy implies an integral approach which takes account of its own broad set of objectives. The focus of such examples is less on agriculture, but makes use of the wide range of measures available.

- A limitation (gradual) decrease of Pillar 1 support might increase the effectiveness of Pillar 2 measures, in particular agri-environmental measures in more-favourable areas. On the other hand, the impact of measures is not only a matter of support levels but is increasingly affected by regulations (c.f. good farming practice) which have a regionally and locally varying influence on farming practices.

- Pluriactivity as a main characteristic of farm households has to be taken into consideration when designing rural development programmes. Innovative ways to address the diversification potential and to respond to emerging societal demand on landscape development have to be improved.

- The concept of multifunctionality is increasingly used as a reference for policy reform. It draws heavily on the rural development paradigm and has clear spatial implications. In designing instruments to the varying needs of regional farming and rural development situations, the place-specific provision of public goods has to be carefully addressed.

- Institutional processes require a long-term involvement and the commitment of all relevant administrative levels. The inclusion of local actors in the bottom-up approach is particularly crucial to the success of rural development initiatives. This involves the need to prioritise facilitation, technical support and capacity building in the RDP to ensure the effective and sustainable use of RDR funds. This process takes time and regional management associations might play a key role in facilitating an advanced understanding of innovative approaches for rural development.

It seems appropriate to anticipate the future policy approach (for the period 2007–2013) in the policies available to the new Member States. This is particularly important to avoid the development of funding structures heavily relying on "phasing out" schemes. However, the current RDP only assigns a limited role to the territorial function and low engagement in rural development measures. The discussion of the general philosophy of policy reform is not only of great influence to the policy shaping in the new Member States, but particularly relevant also for the EU-15 to make sure that they are committed to substantial change.

The spatial distribution of funds underpins the different national and regional priorities for specific instruments. Although these differences partly reflect the regional situations and needs, in many cases they are due to historical allocations of funds to similar measures in the past and a limited range of experience. An exchange of good practice and experiences of RDP implementation between countries would be a key means to learn lessons and support the Member States to improve their use of the wide range of RDR measures. Such networking experiences might also help to redress the balance of funding between regions to more fully match the relative levels of economic, environmental and social needs across the varied territory of rural regions in Europe.

The current reform discussion is heavily led by the issue of shifting resources from Pillar 1 to Pillar 2. Although this implies a reduction of funds allocated to market support, the funds available for the Second Pillar largely are used for accompanying agricultural instruments and generally only available for farmers. Thus it seems hardly an integrated rural development approach. However, the examples provided by regional integrated programmes (like Objective 5b programmes in the 1990s) and local action groups within the LEADER initiative include a host of interesting examples where a particular territorial perspective is taken and strengths and weaknesses of regions are assessed. The instruments of such a policy have a development rationale, proactive rather than compensatory. It is referred increasingly to the LEADER method when future rural development policy is discussed. The mainstreaming of its approach and key elements is therefore regarded as a decisive challenge to substantially change the contents of RDP and increase the spatial dimension. Nevertheless it will be important to take particular care to maintain the innovative character of LEADER-like measures and allow continuation of such activities in future programmes and also participation for non-farming actors.

Analyses of policy application have two more key aspects: the integration of environmental aspects into CAP and rural development policy and the integration of spatial aspects into CAP. As outlined above, there is a clear trade-off between the respective objectives of different policy fields. However, it has been made clear over the recent decade that a more integrative approach is required. This implies the extension of the regulation framework both to these issues and to the programming processes and requirements. Although some of the key principles have been adopted in EU regulations and a series of strategic documents issued there is still a lack in the implementation process to reflect these concerns. The existing application and good practice experience needs an in-depth evaluation which goes far beyond the actual programme evaluation practices, but extends to the more integrative impact analysis on environmental performance and regional outcome. Cross-national discussion and networking would encourage those responsible to take these concerns more seriously in policy formulation.

Chapter 9

9. Conclusions and Policy Proposals

9.1. Territorial impacts of the CAP/RDP

In this study, empirical analysis has been conducted at NUTS3 level using data from a variety of sources, some directly recorded at this level but most requiring derivation from sample and/or higher-level values. The quality of the data is discussed further below, but it is believed that this is the best data available and that our results are robust and reliable, except where caveats are made explicit.

So far, the design and implementation of the CAP has been little touched by the territorial concepts of balanced competitiveness, economic and social cohesion, and polycentricity set out in the ESDP and in the Third Cohesion Report, although it has begun to address the goal of environmental sustainability. Neither have the Agenda 2000 or MTR reforms of the CAP, into Pillar 1 (comprised of market support, mostly non-budgetary, and direct payments), and Pillar 2 (agri-environmental and other "rural development" expenditures), been based on cohesion or other territorial criteria. Even in the implementation of Pillar 2 through the rural development programmes of Member States almost all measures have been horizontal across the whole nation or region, except for areas designated for agri-environmental programmes. The CAP thus remains focused on its own historic objectives, set out in the Treaty of Rome, and its subsequent evolution has reflected other internal and external objectives and pressures.

Simple two-variable correlation analysis suggests that total CAP Pillar 1 support does not support territorial cohesion, with higher levels of CAP expenditure per hectare UAA being associated with more prosperous regions. Direct income payments appear to more strongly support cohesion objectives but are dwarfed by the market price support element of Pillar 1. This may not be surprising, since Pillar 1 has never been claimed to be a cohesion measure. The Rural Development Regulation is a cohesion measure, however, and while our evidence on Pillar 2 is more mixed, expenditure under the RDR does not appear to support cohesion objectives.

The level of total Pillar 1 support was found to be generally higher in more accessible regions, lower in more peripheral regions at all spatial scales (local, meso and EU-level). Multiple regression analysis shows that total Pillar 1 support is strongly associated with a region's average farm business size and land cover indicators. In contrast, Pillar 2 support was found to be higher in more peripheral regions of the community. In this case, multiple regression analysis found higher levels of support tended towards regions with smaller farm sizes while land cover variables were found to be less important explanatory factors. For both types of support, after allowing for these other factors, no statistically significant relationships are observed with GDP per head in NUTS3 regions. In other words, the strong tendency for Pillar 1 support to go to richer regions of the EU-15 may be attributed to their larger farms, their location in the core of Europe, and their farm type.

From the numerical analysis presented in Chapter 5, then, it appears that the CAP has uneven territorial effects across the EU-15 which do not support cohesion objectives, particularly in terms of its Pillar 1. The "rural development" Pillar 2 may in some cases be more consistent with cohesion within countries, but runs counter to EU-wide cohesion in the way it is currently structured.

Our study considered in more detail these impacts both through a number of case studies of the use of measures in different countries and regions and through the CAPRI model of the impact of the MTR proposals. A case study of Irish agricultural and rural development illustrates the kinds of adaptations made by farming households. First, the territorial impacts of agricultural and rural development policies vary with the aims of such policies but are also differentiated according to the resource and structural characteristics of regional economies. Secondly, there is a longer-term, underlying process of agricultural restructuring onto which policies are layered. Policies may cushion the more deleterious impacts of this on farm households (e.g. by supporting incomes), thus slowing the rate of structural change, or "go with the flow" while facilitating desirable adjustments (e.g. by promoting alternative forms of land use). Thirdly, policies may have inconsistent outcomes – as for example when farm price policies have territorial impacts that run counter to cohesion objectives. Finally, it is clear from the Irish case study that in the more commercial farming regions a comprehensive range of agricultural policies and/or agriculture-centred rural development policies will not provide a guarantee of rural demographic viability. In "strong agricultural areas adjusting to restrictions in farm output" without a strong non-farm-based economy, population trends were weaker even than those of marginal agricultural areas. There is a need for greater complementarity between agricultural policy measures and policies for broader regional development focused on the specific conditions of the different regions. In the New Member States too this is crucial.

Turning to agri-environment measures, these were found to contribute to prudent management of and protection of nature and cultural heritage through encouraging a reduction in inputs of inorganic fertilisers, conservation of habitats, and preservation of the cultural landscape. Agri-environment schemes are particularly suited to the encouragement of appropriate land management. The provision of support for organic production, given a high priority in several countries, has the potential to contribute to balanced competitiveness through high quality food production targeted at niche markets. Agri-environment programmes can also make an important indirect contribution to economic and social cohesion through the provision of income support in marginal areas, thus contributing to the retention of rural population.

Even though these measures are usually horizontal, especially in respect of organic production and training, such programmes have been largely identified with environmentally sensitive and extensive farming areas, with the notable exception of Austria where the aim is the "ecologicalisation" of all agricultural activity. It appears that in lowland areas of more intensive farming, regulation through cross-compliance is more effective than agri-environmental measures. Incentives are generally not adequate to encourage participation among more intensive and commercially oriented farmers whilst eligibility criteria are also a barrier to participation. Moreover, the effectiveness of the programme has also been compromised by poor targeting and the continuation of production-linked support policies associated with environmental problems (i.e. support for intensive farming with potentially negative environmental impacts). Finally, and as noted above, agri-environmental measures are used more in the more prosperous regions of northern and western Europe.

A second measure considered was early retirement schemes (ERS), which have been used to achieve both social and structural objectives. Their design (and uptake) has varied by country and depends largely on national objectives. It was concluded that they have been more successful in ensuring the continuation of family farming and population stabilisation than enhancing competitiveness and structural adjustment. However, in the countries with the highest rates of participation (France, Greece and Ireland), the structural effect was little different from that which would have occurred anyway, albeit over a slightly longer time scale. These time gains offered by the ERS are important only in relation to the depopulation problems and the demographic scarcity of farm successors prevailing in LFAs. Within France, Ireland, Norway, Finland and Spain, a distinct spatial pattern of adoption of the ERS exists: the highest levels of adoption were reported in areas of least need (i.e. prosperous farming regions) and where there are higher numbers of young farmers. Population density emerges as an indicator of the regional propensity to early retirement. On this basis, early retirement schemes did not appear to offer benefits in terms of balanced competitiveness, territorial cohesion or sustainable development, except in a very few Less-Favoured Areas (LFAs).

LFA compensatory payments were the next measure considered. The spatial differences of European agriculture are reflected in the application of this scheme. In contrast to what one would expect from a "compensation" measure the application of the scheme is largely correlated to the degree of farm net value added, i.e. higher CAs are applied in more prosperous countries, and in "poorer" countries only a low level of CAs is achieved. The lower commitment of southern Member States is partly due to the prevalence of arable land and permanent cropping in the LFAs of the South (whereas the scheme is largely oriented towards livestock farming) and the focus on modernisation schemes and the improvement of processing and marketing structures. A major reason for this spatial distribution of funds is that the reference level is set at the national level, and not at the European level, such that differences between Member States remain unchanged.

The steady extension of the LFA area since its initiation in the 1970s reflects the political process of defining the border of LFAs, and gives rise to further discussion on the criteria of delimitation and internal differentiation. The review of the intermediate zones proposed by the Commission in July 2004 will address this issue. As the extension has been partly accompanied by an increase of overall grants, at least in some countries, the support level per unit did not fall. The recent changes in the LFA scheme (to an area basis) did not only have an impact on farm management itself but also on farm incomes. In several countries the changes were cushioned by an increase of CA funds and/or a transition period. Finally, LFA payments often underpin high nature value (HNV) farming systems. The existence of HNV farming systems in these areas points to the beneficial role of LFA payments for nature conservation and biodiversity, especially now that these payments are decoupled from livestock numbers. However, these farming patterns are highly threatened by impending marginalisation processes which are particularly relevant for peripheral situations, including regions of the new Member States.

The final measure considered in these case studies was Article 33 and LEADER-type measures. The evaluation studies (of LEADER II and the mid-term evaluation of LEADER+) suggest that such initiatives have a considerable impact on the development of rural regions, although their *budget* is small compared to mainstream programme instruments.

The ex-post evaluation of LEADER II found the programme both efficient and effective. It proved to be *adaptable* to the different socio-economic and governance *contexts* and applicable to the small-scale, area-based activities of rural areas. It could therefore also reach lagging regions and vulnerable rural territories. LEADER activities induced and conveyed responsibility to local partnerships, linking public and private institutions as well as different interests of various local actors to a common strategy. A profound change from a passive to an active attitude could be achieved among many local actors. In countries with a long-standing tradition of pluriactivity, agricultural diversification served as a basic pattern for multi-sectorial strategies, often in combination with rural tourism. A good example for the multi-sectoral approach based on agricultural products and rural tourism is analysed in the Austrian LEADER case study. In some other countries, LEADER projects focused mainly on environmental measures trying to protect and further develop existing natural capital.

LEADER is not an instrument to change local economic structures or revalorise the local economy in a direct way, but rather an instrument to stimulate processes in the local economy so leading to indirect but enduring benefits. Many core projects do preliminary work in activating rural actors, and this is then a stimulus to further economic activities. The potential of LEADER lies especially in the improvement of intangible factors, in raising awareness, in strengthening strategy and cooperation within the region. This often builds the basis for the provision of better services and more competitive products in the longer term.

Following the case studies of these specific Pillar 2 measures, the impacts of the MTR proposals were analysed using output from the CAPRI modelling system developed at the University of Bonn. The modelling system involves physical consistency balances, economic accounting, considerable regional specification (e.g. set-aside rates, direct payment rates, etc.; for non-EU regions, OECD PSE/CSE data are used), and standard micro-economic assumptions. Given the objectives of our study, analysis was restricted to considering the estimated impact of MTR on farm incomes in 2009 relative to their level in the absence of reform. The principal conclusions of this analysis are that farm incomes in the EU-15 (including CAP premiums) are expected to be only marginally affected by the MTR proposals, with changes of more than 5% apparent only in a small number of NUTS3 regions in France (mainly in the south) and in Austria (both show falling incomes) and in some or all of Northern Ireland, Belgium, northern Italy, Denmark and Sweden (all show rising incomes). Analysis found no statistically significant relationship between MTR impacts and cohesion indicators (GDP per head, unemployment rate and population change). Importantly, this suggests that the latest reforms of the CAP will do nothing to remove the existing inconsistencies between the CAP and cohesion policy unless they are accompanied by specific national priorities aimed at regional specific programme implementation.

Returning to the issues raised in Section 1.3, it is apparent that the agro-industrial model still dominates the CAP, and the thinking of those who administer it. Not only does the CAP work against other EU objectives of cohesion, there is a lack of the information necessary for a more integrated, territorial approach, whether for post-productivist or sustainable rural development ends. While some recent reforms have allowed Member States to operate an increasing range of discretionary support measures directed towards territorial priorities, this is still only a very minor part of the CAP. Moreover, this partial renationalisation may be at the expense of EU-wide solidarity and cohesion, as richer countries and poorer countries use different measures with unequal means, unless the financial allocations and the requirements for co-financing are addressed.

9.2. Good practice in rural development

Rural development is a broad concept covering many different perspectives and priorities. Given the vulnerable and often less successful economic performance of rural areas in comparison to urban areas, economic development and viability are core issues for the future. It is now generally understood that a purely sectoral approach is less successful in enhancing and stabilizing a region's performance, but despite this the notion that rural development goals widely overlap with agricultural policy is still characteristic of the CAP. An integrated, territorial approach, sensitive to the diversity of rural circumstances, is needed to ensure regionally balanced development and territorial cohesion.

While tangible factors such as natural and human resources, investment, infrastructure and economic structure have traditionally been seen as the main determinants of differential economic performance, more recent research has highlighted the important role of "less tangible" or "soft" factors, including various kinds of social, cultural, institutional, environmental and local knowledge which constitute the basic capital for regional development. Social capital, especially, has been identified as crucial (Putnam, 1993). A recent EU project on the Dynamics of Rural Areas (DORA) (Bryden and Hart, 2004) has suggested that it is the relationship between the tangible and less tangible resources, and how these interact in the local context, which conditions opportunities and constraints for local development. "It is not so much the tangible resources themselves that matter for economic performance, but the way the local people are able to exploit those available to them" (Bryden and Hart, 2001, p. 45). Thus "less tangible" factors determine the efficiency and effectiveness with which tangible resources are used and are most important in making the difference (Pezzini, 2003, p. 5).

A conclusion emerging consistently from many recent studies, then, is that *social* processes are fundamental to rural development. In this sense, social capital has a vital role in rural development, along with appropriate structures of governance. The role of public policy and development agencies is seen increasingly as to trust, foster and enable local action. This has a number of implications for policy.

The EU RESTRIM project (Arnason *et al.*, 2004) concluded that public policy should therefore support the social processes which are as essential to rural development as "hard" economic intervention (in the same sense that software is as necessary as hardware to computing). In practice this means supporting *rural community development* – understood as an approach to working with and to building the capacity of individuals and groups within their communities. This approach seeks to strengthen communities through enhancing people's confidence, knowledge and skills, and their ability to work together. In the EU, this type of approach has been piloted successfully under the community initiative, LEADER, as noted above, and the Commission has proposed that this is continued and encouraged after 2007 within the single rural development fund.

A number of studies have also suggested that supporting the development of *vertical and horizontal networks* in community action can transcend the dichotomy of endogenous/exogenous development ("bottom-up/top-down"). Issues will arise of where power and control lie in these networks, and of whose problems they are addressing and who benefits, and public bodies and development agencies should be alert to these aspects when offering support and when working with voluntary and community bodies. *Training* of local and regional officials, and others, in the social processes surrounding local development is crucial.

Thirdly, in offering grants and other support, development agencies should prioritise *collective action which is both inclusive and reflexive,* and should support new arenas for interaction. Good networks are inclusive, facilitating collective learning, allowing sharing of success and generating wider social acceptance. In this context, it is notable that most expenditure under the EU Rural Development Regulation has hitherto been targeted largely at individuals rather than collective activities. The RESTRIM research noted the scope for the RDR to be more effective through promoting collective action.

All recent studies have concurred that appropriate *structures of governance* are also essential to facilitate local leadership and innovation. Rural areas and people require strong support from national government and the EU, as well as from regional agencies and the private sector, and it is essential that these set a coherent framework within which participative local development initiatives can flourish. Within such a framework, rural development can be pursued which is locally embedded, socially inclusive and linking social scales. Successful development of this type frees rural areas from stereotypes of backwardness, remoteness and parochialism, and yet allows them to retain control of distinctive and valued cultural and environmental features, with long-term beneficial results. Thus, both the DORA and RESTRIM projects emphasised the importance of effective and open governance, with a positive attitude to small local enterprises and entrepreneurs, and local public institutions with sufficient autonomy to adapt policies and specific measures to assist with the collective needs of local enterprises. Furthermore, open and inclusive "soft" networks are positively related to the mobilisation of entrepreneurial capacity and local initiative.

In the current discussion of rural development, the cultivation of *rural amenities* is often seen as a means of generating new economic opportunities. Rural amenities are strongly associated with specific territorial attributes. Their value stems from the unique features of a given region which cannot be (easily) replaced or exchanged (less-mobile). Thus, it gives the same region a chance to enhance its competitiveness through "cultivating" the place-based social, cultural, and environmental assets. Ideally, this high valorisation from outside the region contributes to strengthening the high value placed on these amenities by local people, which symbolises their distinctive cultural identity. The successful cultivation of rural amenities needs both the regional identification of natural and cultural amenities and favourable structures of decision-making processes in the region. This includes for example local institutions with capacity to recognise the market value and who are able to organise and coordinate supply and promotion of the specific local amenity. The main conclusion from the RESTRIM project, however, is that this is a highly tensioned process that cannot be simply controlled by key development actors: it is important to reflect a plurality of cultural identities and to link this to cultures of everyday life through a broad participative process. Newly constructed regional identities will only succeed in mobilising common efforts towards shared objectives where these supplement and build on multiple local identities.

Some examples of either innovative approaches or representative use of the RDR framework are considered in Chapter 8. Innovations in RDP across Europe have occurred both at the level of programme design and resourcing, and at the level of individual projects and initiatives. Many examples demonstrated flexibility and tailoring of measures to meet local circumstances and potential. These included:

- Differentiation of compensatory allowances for LFA in Austria
- Ireland Rural Environment Protection Scheme (REPS)
- CTE (contrat territorial d'exploitation), France
- "Cheese Route Bregenzerwald", LEADER, Austria
- Rural Tourism in Italy
- PRODER Andalucia, Spain
- POMO and POMO+, Finland.

The achievement of ESDP objectives relating to prudent management of resources depends on effective integration of environmental measures within the CAP. Possible mechanisms for integration include cross-compliance and the verifiable environmental standards required for certain measures under Pillar 2 as well as a significant expansion of Pillar 2 measures. In order to raise effectiveness, Member States should define measures with specific environmental objectives rather than focusing on agricultural practices. The potential benefits of an integrated approach to EU structural and regional policy instruments are also supported by the conclusions from the Schramek *et al.* (1999) report which recommended improved integration of existing structural and regional policy and instruments such as the LFA scheme and LEADER with agri-environmental policy.

In terms of LFAs, Member States have developed nationally shaped instruments which are particularly adapted to their specific situations and priorities. We can discern, therefore, a great variety in the application of this instrument across the EU. Only in some countries has a detailed differentiation of production difficulties within the areas been implemented (e.g. Austria). Elsewhere, the level of support fails to reflect production difficulties. As a result the measure is criticised, in particular with regard to under-/overcompensation, local/regional equity, and lack of international consistency of support levels/income levels. The instrument should address more directly these objectives by differentiating payments according to, and including criteria for the measurement of, production difficulties. Administration costs of such systems are less high than might be expected since new technologies (e.g. aerial photogrammetry and remote sensing, GIS applications) allow for a highly advanced (automatically updated) control framework which may be used in conjunction with requirements for other CAP payments. More difficult policy choices have to be made as regards social as well as environmental questions, e.g. the desirability of maintaining traditional or at least local farm management (instead of incomers or "remote" management), and the "problems" of dealing with micro holdings maintained privately for seasonal and/or recreational use.

Finally, the LEADER Community Initiative is one of the four remaining EU CIs for the period 2000–2006, but has a very limited budget (€2.02m), compared to the overall Structural Funds and CAP budgets. Nevertheless it is the programme which is most closely related to the concept of integrated rural development, and provides a multitude of good/bad examples of rural development under different contexts. Moreover this pilot programme has had a crucial impact on the political discourse and on the discussion of regional development in peripheral areas. Beyond the economic sphere the programme is important for other spheres of rural life and policy, due to its multi-sectoral and integrative character.

LEADER provides a flexible programme structure which has to be adapted to the context of the rural regions, and has achieved interesting results for small-scale regional development. Numerous case studies (beyond those carried out under the ESPON programme) elaborate on the starting period, the difficulties and outcome of initiatives. Some of them also underline the requirements for the successful application and institutional prerequisites, including the following characteristics of action-centred networks: flat, flexible organisational structures involving teamwork and partnership; equality of relationships among relevant stakeholders; vision and value-driven leadership, and emphasis on participation and organisational learning. The core of the programme is the emphasis on the multi-sector approach which requires a high commitment by participants to overcome institutional and deeply entrenched personal difficulties with regard to cooperative activities and new ways of organisation at the local level. This discussion has turned out to be very important for the discussion of regional governances.

During the LEADER programme period evidence has emerged of an increasing level of rural development "know-how" and an improved capacity of partnerships to deliver programmes for rural development. LAGs no longer see themselves mainly as a provider of local funding on a project-to-project basis, which often resulted in a "scatter-gun" approach to development. This change to a programme-driven approach has enabled LAGs to manage and target resources in a more effective and pro-active manner. Nevertheless, in some respects participation remains unsatisfactory. In particular, different groups of society are underrepresented and LAG strategies reflect local power relations in the LEADER areas. An enlargement of the groups addressed and integrated in the process is one of the objectives of the LEADER+ (e.g. stronger participation of women, young people etc.) and would further enhance the effectiveness of this approach.

9.3. Synthesis

The principal conclusion from this ESPON project is that in aggregate the CAP works against the ESDP objectives of balanced territorial development, and does not support the objectives of economic and social cohesion. Moreover, in terms of polycentricity at the EU level, Pillar 1 of the CAP appears to favour core areas more than it assists the periphery of Europe, and at a local level CAP favours the more accessible areas. In recent years the CAP has undergone a series of reforms. Some of these have begun to ameliorate these conflicts of objectives. For example, direct income payments tend to be distributed in a manner more consistent with cohesion than market price support. Similarly, higher levels of Pillar 2 payments are associated with more peripheral regions of the EU than is the case with Pillar 1 support. Nevertheless, there is considerable scope for both Member States and the Commission to make the CAP more consistent with the objectives of the ESDP. It is encouraging that senior officials of DG Agriculture have placed importance on "the difficult question of how we can centre our policy more around the territorial instead of the sectoral, i.e. agricultural, dimension of rural development" (Ahner, 2004, p. 12). This is reflected to some limited extent in the announcement that 7% of the RDR will be devoted to LEADER-type measures from 2007.

The scientific evidence suggests that there is scope to amend Pillar 2 to favour cohesion, and that this holds out the best potential for amending agricultural and rural development policy and policy instruments to support territorial cohesion and the ESDP. We concur with the conclusions of Dwyer *et al.* (2002) that "the RDR is an innovative tool with considerable potential to support sustainable rural development throughout Europe, particularly in promoting a more integrated and multifunctional approach to rural land management, environmental integration and economic and community development," but that this potential is not currently being realised. "Planning and implementation of the RDR and SAPARD do not reflect the ambitions of the Commission's objectives" for the Second Pillar, because of: "lack of time for planning; complex administrative procedures; inadequate funding; and limited incentives for countries to re-think and re-design existing policies to reflect fully the scope of this new instrument and its requirements." Moreover, the Second Pillar is still focused mainly on agricultural producers rather than on territorial rural development, and this will remain so under the revised RDR for 2007–2013.

9.4. Policy proposals

It may be helpful to begin by summarising the main conclusions of the Salzburg Conference organised by the European Commission in November 2003. There was consensus around three broad objectives (see http://europe.eu.int/comm/ agriculture/ecrd2003/):

- *a competitive farming sector*: Sustainable economic growth in farming must come increasingly through diversification, innovation and value added products;
- *managing the land for future generations*: managing the farmed environment and forests should serve to preserve and enhance the natural landscape and Europe's cultural heritage; and
- *a living countryside*: investment in the broader rural economy is vital to increase the attractiveness of rural areas, promote sustainable growth and create new employment opportunities through diversification.

It should be noted that the first of these objectives is inherently non-spatial, except insofar as the agri-food sector (rather than policymakers, who can only reinforce commercial trends) can find and add value to local and regional farm output. There is no obvious reason why all EU regions should be able to operate effectively in an increasingly competitive and widespread market, and it should not be expected that agriculture, even if diversified or innovated, can in future support previous levels of farm occupiers and incomes. In regions which "lag behind" despite best efforts, policy attention directed at territorial cohesion must shift even further towards alternative sources of economic activity and income. Objectives 2 and 3 above are more capable of direct territorial interpretation in policy terms, but, as demonstrated in a previous section, experience shows that national and other factors are unlikely to promote EU-wide cohesion effectively unless careful account is taken of relative territorial capacities and resources.

The Salzburg conference also concluded that rural development policy should apply in all rural areas of the enlarged EU; and that rural development policy must serve the needs of broader society in rural areas and contribute to cohesion. In other words, rural development should be more than just a sectoral approach linked to agriculture. It clearly has an important territorial dimension.

The EU Commission has taken these conclusions, along with a number of evaluation studies, as a main point of departure in reviewing its rural development policy. In particular, it has proposed grouping the different measures in the RDR around the three core priorities suggested by the Salzburg conference, along with a fourth axis of LEADER-type measures. Such an approach envisages substantial flexibility for Member States and regions in the implementation of these measures, while at the same time promoting EU strategy by prescribing a minimum proportion of the budget to be devoted to each heading. Thus, at least 15% of each country's national envelope has to be spent on Axis 1 (Improving competitiveness of farming and forestry), at least 25% on Axis 2 (Environment and land management); and at least 15% on Axis 3 (Improving quality of life and diversification), and in addition at least 7% on a new Axis 4 (LEADER). Moreover, the RDR budget would be increased substantially to €13bn per annum (EC, 2004c).

Earlier a senior official had suggested that as much as 30% might be earmarked for mainstreaming LEADER (Courades, 2004), with permanent support structures for capacity-building, networking and vertical and horizontal coordination. On the basis of our scientific conclusions, we would also recommend larger spending on such a LEADER-type approach if territorial cohesion is to be pursued. Nevertheless, the more gradualist proposals will allow the LEADER model to be applied on a wider scale by the Member States who wish to do so, "while for the EU as a whole continuation and consolidation of the LEADER approach will be safeguarded" (EC, 2004c). The Commission argues that its proposals "will ensure better focus on EU priorities, and will improve complementarity with other EU policies (e.g. cohesion and environment)." Our findings support this claim.

9.4.1   Specific proposals

We would propose, first of all, that the Pillar 2 budget should be increased progressively, as anticipated in the Agenda 2000 and MTR agreements and in the Commission's proposals for the RDR 2007–2013. This might be achieved either through continuing increases in the rate of compulsory modulation (which would attract/require match funding) or preferably through the more substantial realignment of EAGGF towards Pillar 2. This is desirable because the RDR incorporates cohesion objectives, in contrast to Pillar 1. This proposal follows directly from our conclusion that Pillar 2 offers the best potential for amending agricultural and rural development policy to support territorial cohesion and other ESDP objectives. The proposals for the RDR 2007–2013 represent a significant step in this direction, and the more quickly support is transferred from Pillar 1 to Pillar 2 the more consistent the CAP will become with cohesion objectives. Moreover, as the Buckwell Report (Buckwell *et al.*, 1997) argued, the expenditure of funds under the CAP will be more defensible if they are directed towards "public goods" such as the cultural and natural heritage, environmental benefits and sustainable rural communities.

We recommend that the new Rural Development Regulation 2007–2013 should contain a broader range of permitted measures under the four proposed axes, building on the lessons from LEADER and Objective 5b by including more measures which address sustainable rural development beyond the agriculture sector and which have a territorial dimension. Encouragement should be given to innovation. The revised RDR 2007–2013 strikes a balance between pursuing an overall EU strategy for rural development and greater subsidiarity, allowing RDP to be tailored more appropriately to the diversity of territorial needs across rural Europe, but most measures are still to be sectoral rather than territorial. More measures should be open to non-farmers and build on the lessons of LEADER, Objective 5b and DORA, as implied by "Mainstreaming LEADER" and the Salzburg conclusions.

It is important these territorial measures include supporting rural community development – understood as an approach to working with and to building the capacity of individuals and groups within their communities. To this end, in offering grants and other support, local development agencies should prioritise collective action which is both inclusive and reflexive, and should support new arenas for interaction and collective learning.

We recommend that the Commission keeps under review the rates of co-financing in the convergence countries, as there is evidence that the difficulties of match funding may have led both to lower levels of RDR expenditure and to a distorted composition of RDR spending in the poorer countries and regions. The Commission's proposals to allow significantly higher rates of EU co-financing in the convergence countries from 2007–2013 are welcomed.

We also point out that consistency with cohesion objectives would be improved through allocation of the RDR budget to Member States according to criteria of relative needs for rural development and environmental management, as proposed by the Commission (CEC, 2002). A recent paper by Mantino (2003) has illustrated a variety of ways in which this might be achieved at a regional level, using weighted criteria suggested by the Commission in the first draft of the MTR proposal (UAA, agricultural employment, GDP/head) and already used for SAPARD allocations in the then Candidate Countries, as well as various environmental criteria (Natura 2000 sites, protected areas, organically farmed area).

Turning to Pillar 1, it is likely that there will be further revisions of the Market Price Support arrangements as a result of the currently ongoing WTO negotiations. The 31 July 2004 agreement covers reductions in export subsidies, border protection and trade-distorting domestic support to agriculture, and, once actual modalities (formulas) and numerical values have been agreed in future talks, should lead to further lowering of EU market prices, especially in products (e.g. sugar, beef) which have retained high border protection. The more that WTO negotiations result in reductions in Pillar 1 Market Price Support, through reductions in border protection and a convergence of EU prices with world prices, the greater the resulting consistency of the CAP with cohesion objectives. As our scientific results have shown conclusively, the Market Price Support element dominates the CAP and benefits overwhelmingly the richer, core regions at the expense of the poorer, declining and more peripheral parts of the EU. The gradual reduction of this element of CAP support is fundamental to any reorientation of the CAP towards cohesion objectives.

In relation to direct Single Farm Payments, it is suggested that the Commission explore models through which these might be modulated more progressively in richer regions of the EU, for example through relating rates of modulation to farm business size. Voluntary modulation could previously be applied in this way (as was done briefly in France) with a positive territorial impact, and this would be worthy of further investigation.

This approach will require greater harmonisation with regional policy, and will also require attention to be paid to appropriate institutional structures for multi-level governance (see below).

9.4.2 Institutional proposals

The ESDP challenges us to move towards a holistic and integrated approach to both the understanding and the implementation of sustainable development. The need for such an approach appears to be greatest in the poorest regions of the Community, eligible for Objective 1, where a "very high degree of sectorialisation" prevails (Robert *et al.*, 2001), but is also required elsewhere. Local development strategies, as proposed by the Commission in Axes 3 and 4 of RDR 2007-13, offer a means of integrating the approach to policy delivery and combining various instruments and funding streams for maximum effectiveness. Such strategies should seek horizontally integrated solutions combining actions in different sectors (economic, social, environmental). It is also imperative, however, that they should achieve vertical integration between local, regional, national and international funding and actors

Those operating at EU, national or regional level must play an important role in setting a coherent framework within which local development initiatives can best add value to top-down approaches. In particular, they should secure coordination at the highest levels where mainstream policies and strategies are formulated, so that top-down policies can effectively be integrated at local level by local development agencies and so that vertical integration can be achieved between local, regional and national policies. In addition, there must be a suitable mechanism for effective coordination of local development programmes, to avoid duplication or conflict. It will also be helpful to encourage a horizontal learning process between regions and between local actors in different territories.

The issue of appropriate institutional structures of multi-level governance is therefore of considerable importance, and we offer the following recommendations:

- The integrated development of land use, linkage to other local sectors and the creative development of region-specific programmes (as outlined in Chapter 8) are necessary to enhance the cohesion aspects of the CAP. Such an approach would require stronger regional programming for specific rural development measures, and the opening of "Rural Development Programmes" to all the rural population, rather than only or mainly to farmers.

- We echo the conclusions and recommendations of Robert *et al.* (2001) who argued for institutional readjustments at Community, national and regional levels to allow the establishment of a correct balance between the various administrative levels associated with the sectoral and territorial policies affecting rural areas; and for greater flexibility of operational programmes and Community Initiatives, and even certain aspects of the CAP, to take account of the differentiated countryside; input into strategic objectives and visioning from local communities; and partnership arrangements at the operational level which provide the mechanisms for integration.

- Institutional processes require a long-term involvement and the commitment of all relevant administrative levels and Directorates.

- The inclusion of local actors in a bottom-up approach is particularly crucial to the success of rural development initiatives. This involves the need to prioritise facilitation, technical support and capacity building in the RDP to ensure the effective and sustainable use of RDR funds. Again, this process takes time and regional management associations might play a key role in facilitating an advanced understanding of innovative approaches for rural development.

- An exchange of good practice and experiences of RDP implementation between countries would be a key means to learn lessons and support the New Member States in their use of the wide range of RDR measures. Such networking experiences might also help to redress the balance of funding between regions to more fully reflect levels of economic, environmental and social needs across the varied territory of rural regions in Europe.

These suggestions regarding institutional issues are made in most studies and command a broad support in the literature. A fundamental question then is why so little progress has been made and what might be done to promote change. Our final recommendation, then, is that both the Commission and Member States start reviews of their institutional arrangements for rural development and agriculture, encompassing broad consultation and debate, and leading to firm proposals.

9.5. Data requirements

The availability of detailed territorial data on agriculture across Europe is surprisingly poor, given the huge extent of agricultural data collection and the bureaucratic burden on farmers. Very little data relating to agriculture are available at NUTS3 level from Eurostat, DG Regio or DG Agriculture, and where they do exist up to 91% of data are missing. DG Agriculture reported that they have no information on CAP expenditure below national level other than Farm Accountancy Data Network sample data, which shows support received rather than expenditure.

We have therefore encountered persistent difficulties in capturing territorial specific information on CAP performance in general, and on separation of different CAP instruments, despite the huge routine surveillance of farmers. It is especially surprising that DG Agriculture apparently has no systematic information on the regional pattern of CAP expenditure. The only indicator from the REGIO dataset widely available at NUTS3 level relating to agriculture is employment in agriculture, forestry and fishing (derived from the Regional Accounts), and missing data is a problem for this and many other variables. Similarly the FADN dataset only provides data at NUTS2 or NUTS1 level, and sometimes in non-standard areas. We have made the best of the available data, using reliable national and OECD data to supplement EU sources and to derive robust NUTS3 estimates from sample and/or higher-level values. Nevertheless, data should be provided to the Commission and published at NUTS3 or even NUTS5 level.

One conclusion is that information on CAP expenditure and implementation at regional level is poorly developed, and support to overcome this information gap is limited. As the territorial dimension becomes integrated into rural policy, it will be very important to support policy-making in future through improving the database so as to enable Europe-wide territorial analysis. This will require administration of the CAP instruments to take into account the regional and territorial dimension.

At the same time the lack of useful regional information also reveals a lack of understanding of, or commitment to, the territorial relevance of the CAP amongst officials. Instead, it appears that most tend to think predominantly or only of linkages as upstream and downstream (i.e. within the farm-supply and food chains), rather than as existing in space. A cultural change amongst officials (reinforced by revised policy objectives and criteria) is needed if they are to address the territorial dimension of agricultural and rural development policy in future.

Specific data requirements to permit future monitoring and analysis include:

- CAP expenditure by policy measure at NUTS3 level;
- Outputs of principal commodities, annually, at NUTS3 level;
- CORINE land cover change estimates at NUTS3 level;
- Farm numbers, farm workforce, and subsidy receipts at NUTS3 level;
- Level and composition of farm household incomes at NUTS3 level;
- Proportion of each NUTS3 area designated under environmental legislation;
- Proportion of each NUTS3 area designated under Structural Funds and LFA;
- Proportion of each NUTS3 area covered by LEADER programmes.

Additionally, much basic data available for the EU-15 is lacking for the NMSs, e.g. population change at the NUTS3 level.

9.6. Further research

Within the time constraints of this project we have been unable to work with estimates of the impacts of the MTR as agreed in June 2003 and as implemented by Member States. Indeed, many MS are still deciding how to implement the MTR. Instead we have made use of the CAPRI model estimates of the impacts of the Commission's MTR proposals, as explained in section 6.2. However, the CAPRI team continue to revise their modelling and to extend their model to cover the NMS.

There would be considerable value in updating and extending to the NMS our analysis of the impacts of the MTR at NUTS3 level, as Member States agree how precisely they will implement the MTR (e.g. on what basis Single Farm Payments will be made) and as further outputs from the CAPRI model become available. Variation in the basis of SFPs may lead to much greater territorial differentiation in the impact of the MTR reforms on environment, farm incomes and practices.

Similarly, within the next few months it will be possible to use the latest CORINE data to incorporate the analysis of changes in land cover at NUTS3 level.

In this book, we have necessarily focused on policy instruments and their impact, while recognising that many other factors also give rise to territorial divergence and issues of spatial development in rural areas. The case study of household adjustment strategies and trends in Ireland was particularly useful in examining this broader picture, and it is suggested that this might usefully be supplemented by similar, additional country studies drawn especially from the Mediterranean countries and from the new Member States.

Two particular issues may also be noted as important for future research. The first of these concerns territorial implications of changes in the food supply chain. Major changes are continuing in the dominant agro-industrial food system, with increasing concentration of power amongst global food corporations and retailers, and these changes might be expected to have significant territorial implications. At the same time, there are counter-movements towards the "relocalisation" of food, through short supply chains and alternative food networks (e.g. "slow food"). Very little work has been undertaken to examine the implications of these tendencies for spatial development. The second issue we wish to highlight concerns local labour market problems and particularly the growing reliance of the agricultural industry on casual labour, often provided by immigrants. It may be worthwhile to explore the spatial aspects of this tendency, and of a growing vulnerability of certain rural economies and territories to changes in migration patterns and regulations.

Finally, we recommend a Futures Study specifically attempting a foresight analysis of the rural areas of the EU-27. Under conditions of rapid change, scenario analysis, horizon scanning and foresight exercises have become a common tool in the visioning of rural futures (Newby, 2004), allowing recognition of strategic choices and pinpointing crucial data requirements. In the context of multi-level governance, differentiated tendencies of rural territories and divergent policy traditions, such a study could play an important role in promoting a more coherent and integrated policy response from the EU, Member States and regional and local actors.

Abbreviations and Glossary

ACP	African, Caribbean and Pacific (countries)
Acquis Communautaire	The body of EU law, regulations and procedures
ACs	Accession Countries
AEP	Agri-Environmental Programme
Agenda 21	A comprehensive plan of action to be taken globally, nationally and locally by UN organizations, governments, and major groups in every area in which humans impact on the environment
AgraCEAS Consulting	Specialist agri-food consultancy established in Brussels in 1973
ANOVA	Analysis of variance
ARMA	Agency for Reconstruction and Modernisation of Agriculture (Poland)
AsPIRE	Aspatial Peripherality, Innovation and the Rural Economy: a comparative research project involving partners in Greece, Scotland, Spain, Ireland, Germany and Finland
AWU	Annual Work Unit
BABF	Bundesanstalt für Bergbauernfragen (Federal Institute for Less Favoured and Mountainous Areas), Austria
BMLFUW	Bundesministerium für Land- und Forstwirtschaft, Umwelt und Wasserwirtschaft (Federal Ministry for Land and Forestry, Environment and Water Management), Austria
CAD	Contrat d'agriculture durable
CAP	Common Agricultural Policy
CAPRI	Common Agricultural Policy Regional Impact (model)
CAP-STRAT	Common Agricultural Policy Strategy for Regions, Agriculture and Trade (EU research project QLTR-200-00394)
CEB	County Enterprise Board (Ireland)
CEEC	Central and Eastern European Country
CI	Community Initiative
CORINE	A pan-European inventory of biophysical land cover, using a 44-class nomenclature

CPB	Centre for Economic Policy Analysis in The Hague
CTE	Contrat territorial d'exploitation
CWFS	Centre of World Food Studies in Amsterdam
DHI	Disposable Household Income
DORA	Dynamics of Rural Areas
DPs	Direct Payments
EAGGF	European Agricultural Guidance and Guarantee Fund
EEA	European Environment Agency
EFTA	European Fair Trade Association
EQUAL	One of the Community Initiative Programmes, designed to promote equal opportunities under five headings: employability, entrepreneurship, adaptability, equal opportunities for men and women and asylum seekers
ERDF	European Regional Development Fund
ERS	Farmers' Early Retirement Scheme
ESDP	European Spatial Development Perspective
ESIM	European Simulation Model
ESPON	European Spatial Planning Observation Network
ESUs	European Size Units
EU	European Union
EU-15	All "old" Member States
EU-25	All EU-15 + N10
EU-27	All EU-15 + N12
EU-27+2	All EU-27 + CH, NO
EUROFARM	A database containing data in the form of standard tables from the Farm Structure Survey
FADN	Farm Accountancy Data Network
FAIR	Fourth Framework specific RTD programme "Agriculture and Fisheries"
FAO	Food and Agriculture Organization of the United Nations
FAPRI	Food and Agricultural Policy Research Institute
FDI	Foreign Direct Investment
FEOGA	Fonds Europeen d'Orientation et de Garantie Agricole; or the Guidance Section of the European Union's Agriculture Fund
FIFG	Financial Fund for Fisheries Guidance
FNVA	Farm Net Value Added
FUA	Functional Urban Area
GAO	General Accounting Office
GDP	Gross Domestic Product
GFP	Good farming practice

GHG	Greenhouse Gases
GISCO	Geographical Information System for the Commission: the database attempts to cover the common interest of the European Commission services in spatial data
GTAP AGE	Global Trade Analysis Project Applied General Equilibrium (model)
GVA index	Gross Value Added index
HBS	Household Budget Survey
HNV	High nature value
I.V.D.	Indemnité Viagére de Départ
IDARA	Integrated Development of Agriculture and Rural Areas in CEE countries
INEA	Istituto Nazionale di Economia Agraria
INTERREG III	Community Initiative which aims to stimulate interregional cooperation in the EU between 2000–2006.
ISDEMA	Innovative Structures for the Sustainable Development of Mountainous Areas (project)
ISPA	Instrument for Structural Policies for Pre-Accession
LAGs	Local Action Groups
LEADER	Liaisons Entre Actions de Développment de l'Economie Rurale
LFA	Less-Favoured Area
LIFE	The Financial Instrument for the Environment, introduced in 1992; one of the spearheads of the European Union's environmental policy
LTP	Long-Term Perspective
MPS	Market price support
MS	Member States: Austria (A), Belgium (B), Denmark (DK), Finland (FIN), France (F), Germany (DE), Greece (GR), Ireland (IRL), Italy (I), Luxembourg (L), Portugal (P), Spain (E), Sweden (S), The Netherlands (NL), United Kingdom (UK), Cyprus (CY), Czech Republic (CZ), Estonia (EE), Hungary (HU), Latvia (LV), Lithuania (LT), Poland (PL), Slovakia (SV), Slovenia (Sl)
MTR	Mid Term Review
N10	All 10 new Member States: see NMS
N12	All N10 + Bulgaria (BG), Romania (RO)
NACARD	National Agricultural Advisory Centres
NATURA 2000	Network for the *in situ* management and conservation of Europe's most remarkable fauna and flora species and habitats, supported by LIFE

NFS	National Farm Survey (UK)
NMS	New Member State. The new Member States in 2004 were: Cyprus (CY), Czech Republic (CZ), Estonia (EE), Hungary (HU), Latvia (LV), Lithuania (LT), Malta (MT), Poland (PL), Slovakia (SK), Slovenia (SI). Candidate countries: Bulgaria (BG), Romania (RO). Others: Norway (NO), Switzerland (CH)
NUTS	Nomenclature des Unités Territoriales Statistiques, or Nomenclature of Territorial Units for Statistics
NVA	Net Value Added
Objective 1	Objective 1 of the Structural Funds is the main priority of the European Union's cohesion policy. In accordance with the treaty, the Union works to "promote harmonious development" and aims particularly to "narrow the gap between the development levels of the various regions".
Objective 2	Objective 2 of the Structural Funds aims to revitalise all areas facing structural difficulties, whether industrial, rural, urban or dependent on fisheries
OECD	Organisation for Economic Cooperation and Development
PHARE	The Phare programme is one of the three pre-accession instruments financed by the European Union to assist the applicant countries of Central and Eastern Europe in their preparations for joining the European Union
POMO, POMO+	Rural programme for local initiative in Finland for the development of rural areas and the archipelago through planning and implementing, in line with the LEADER programme, local strategies and measures coming from the grass roots level. The objective is to create new lines of operation in the rural areas and promote local initiatives
PRIDE	People and Resources Identification for Distributed Environments: a project intended to develop a broker service to support the identification and delivery of information services over the Global Information Infrastructure
PRODER	Programa Operativo de Desarrollo y Diversificación Económica de Zonas Rurales, a set of programmes for endogenous rural development implemented in Spain

PSE/CSE	Producer Support Estimate/Consumer Support Estimate: database produced by the OECD
RDP	Rural Development Policy
RDR	Rural Development Regulation
REGIO	Eurostat's Regional Statistics database
REPS	Rural Environmental Protection Scheme
RESTRIM	Restructuring in Marginal Areas
RICAP	Regional Impact of the Common Agricultural Policy
RUREMPLO	FAIR research project "Agriculture and employment in the rural regions of the EU" (CT96 1766), analysing the development of employment in the rural regions of the EU.
SAPARD	Special Action for Pre-Accession measures for Agriculture and Rural Development
SFP	Single Farm Payment
SME	Small or medium-size enterprise
SMEs	Small and medium-scale enterprises
SPESP	Study Programme on European Spatial Planning
SPP	Special Preparatory Programme
STREP	Specific Targeted Research Project
SWOT	Strengths, Weaknesses, Opportunities, Threats
Teagasc	The Agricultural Research and Advisory Authority, Dublin
TEN	Trans-European Network
UAA	Utilisable Agricultural Area
VAT	Value Added Tax
WTO	World Trade Organization

References and Further Reading

Abrahams, R. (1991) *A place of their own: family farming in eastern Finland*. Cambridge: Cambridge University Press.

AFCon Management Consultants in association with University College Cork (2003) *CAP Rural Development Plan 2000-2006: Mid-Term Evaluation*. http://www.agriculture.gov.ie

Agra CEAS Consulting (2003) Ex-Post Evaluation of Measures Under Regulation (EC) "no." 950/97 on Improving the Efficiency of Agricultural Structures, Final report, The European Commission Directorate-General for Agriculture, www.ceasc.com

Ahner, D. (2004) '*Rural Development and the New Financial Perspectives*,' paper to EU conference on Improving Living Conditions and Quality of Life in Rural Europe, Westport, Ireland, 1 June 2004.

Allaire, G. and Daucé, P. (1995) *Etude du Dispositif De Préretraite en Agriculture: Rapport Général*. INRA Dijon et Toulouse, ENESAD. (in French).

Allaire, G. and Daucé, P. (1996) The farmer early retirement scheme 1992-1994: first balance and structural impacts. *Economie rurale*, 232, 3–12. (in French).

Antikainen, J., Groth, N.B., Meijers, E., Waterhout, B., Zonneveld, W., Cattan, N., Baudelle, G., Baudet-Michel, S., Berroir, S., Bretagnolle, A., Buxeda, C., Dumas, E., Guérois, M., Hamez, G., Lesecq, G., Saint-Julien, T., Sanders, L., Davoudi, S., Strange, I., Wishardt, M., Spiekermann, K. and Wegener, M. (2003) The role, specific situation and potentials of urban areas as nodes in a polycentric development. *ESPON Project 1.1.1, Second Interim Report, March 2003*

Arnalte, E. (2002) Ajuste estructural y cambios en los modelos productivos de la agricultura española. *Agricultura y Sociedad en el cambio de siglo* (ed. C. Gómez Benito and J.J González), 391-426. Madrid: McGraw Hill.

Arnason, A., Lee, J. and Shucksmith, M. (eds) (2004) *RESTRIM: Restructuring in Marginal Rural Areas: the role of social capital in rural development*, Final Report to EU, Arkleton Institute for Rural Development Research, University of Aberdeen.

Arzeni, A., Esposti, R. and Sotte, F. (eds) (2002) *European Policy Experiences with Rural Development*. Kiel KG: European Association of Agricultural Economists, Wissenschaftverlag Vauk.

Asamer, M. and Lukesch, R. (2000) *Actors, Institutions and Attitudes to Rural Development*. The Austrian National Report, The Nature of Rural Development, research report to the World Wide Fund for Nature and the Statutory Countryside Agencies of Great Britain, Graz and Hirzenriegl.

Baldock, D. (2003) Natural amenities and sustainable rural development policies for integration? *The Future of Rural Policy*. Conference proceedings, OECD Headquarters, Paris.

Baldock, D. and Tar, F. (2002) Agri-environment Issues in Central and Eastern Europe – Some Implications of CAP Enlargement. London: Institute for European Environmental Policy.

Baldock, D., Beaufoy, G., Brouwer, F. and Godeschalk, F. (1996) *Farming at the Margins, Abandonment or Redeployment of Agricultural Land in Europe*. London and The Hague: IEEP and LEI-DLO.

Baldock, D., Dwyer, J., Lowe, P., Petersen, J.-P. and Ward, N. (2001) *The Nature of Rural Development: Towards a Sustainable Integrated Rural Policy In Europe*, synthesis report, IEEP, London.

Baldock, D. and Dwyer, J. (with Sumpsi Vinas, J.M.) (2002) *Environmental Integration and the CAP*, report to the European Commission, DG Agriculture. London: Institute for European Environmental Policy.

Baltas, N. (1997) The Restructured CAP and the Periphery of the EU. *Food Policy* 22(4), 329–343.

Bardaji, I., Atance, I. and Tio, C. (2001) Des objectifs territoriaux de la politique d'elevage commune: études de cas espagnols. *Revue de L'Economie Meridionale* 49(193), 25–41.

Baum, S., Trapp, C. and Weingarten, P. (2004) Typology of rural areas in the CEEC new Member States, paper at the 87th EAAE-seminar "Assessing rural development policies of the CAP", 21–23 April, Wien.

Bazin, G. and Roux, B. (eds) (1992) Les Facteurs de Résistance à la Marginalisation dans les Zones de Montagne et Défavorisées Méditerranéennes Communautaires. Paris: MEDEF-network.

Bazin, G. (1999) *La Politique De La Montagne, Rapport D'évaluation*, volume 1+2. Paris: La documentation Française.

BBR (Bundesamt für Bauwesen und Raumordnung) (2001) *Study Programme on European Spatial Planning*. Final Report (Study 103.2). Bonn, (http://www.nordregio.se/spespn/welcome.htm).

Beaufoy, G., Baldock, D. and Clark, J. (1994) *The Nature of Farming: Low Intensity Farming Systems in Nine European Countries*. London: Institute for European Environmental Policy.

Bengs, C. and Zonneveld, W. (2002) The European Discourse on Urban-Rural Relationships: A New Policy and Research Agenda. *Built Environment* 28(4), 278–289.

Bennett, H. (2003) Future Policies for Rural Europe 2006 and beyond – long term support to rural areas in an expanding Europe, conference report, Land Use Policy Group and Institute for European Environmental Policy, London.

Bika, Z. (2004) *Cluster Analysis of NUTS3 Regions*. Aberdeen: University of Aberdeen..

Blanc, M. and Perrier-Cornet, P. (1993) Farm Transfer and Farm Entry in the European Community, *Sociologia Ruralis*. Volume XXXIII(3/4), 319–335.

BMLFUW (Bundesministerium für Land- und Forstwirtschaft, Umwelt- und Wasserwirtschaft) (2000): Grüner Bericht 1999, Wien, p.123ff.

BMLFUW (Bundesministerium für Land- und Forstwirtschaft, Umwelt und Wasserwirtschaft) (2003) *Halbzeitbewertung des LEADER+ Programms Österreich 2000-2006*, Endbericht, Dezember 2003, Wien.

BMVBW (Bundesministerium für Verkehr, Bau- und Wohnungswesen) (2004) *EU informal meeting on territorial cohesion: Presidency conclusions*, Rotterdam 29 November 2004. http://www.bmvbw.de/ Anlage22208/Presidency-conclusions-on-territorial-cohesion.pdf

Bowler, I.R. (1985) *Agriculture under the Common Agricultural Policy*, Manchester University Press.

Bowler, I., Clark, G. and Ilbery, B. (1995) Sustaining farm businesses in the Less Favoured Areas of the European Union, *The Regional Dimension in Agricultural Economics and Policies*, proceedings of the 40th seminar of European Association of Agricultural Economists (EAAE), 26–28 June, Ancona, (ed. F. Sotte), pp. 109–120.

Brangeon, J.L., Jegouzo, G. and Quiqu, M. (1996) Early retirement scheme and low income in the agricultural sector, *Economie rurale* 232, 13–19. (in French).

Britz, W. (ed.) (2004) CAPSTRAT. Common Agricultural Policy Strategy for Regions, Agriculture and Trade. Final report (QLTR-2000-00394), Bonn.

Bröcker *et al.* (2004) *Territorial Impact of EU Transport and TEN Policies*, final report of ESPON Project 2.1.1. Institute of Regional Research, University of Kiel. http://www.espon.lu

Brouwer, F. and Lowe, P. (1998) Cap Reform and the Environment, *CAP and the Rural Environment in Transition: A Panorama of National Perspectives*, (eds F. Brouwer and P. Lowe), 13–38. Wageningen: Wageningen Pers.

Brun, A. and Fuller, A.M. (1992) Farm Family Pluriactivity in Western Europe /Pluriactivité des menages d'agriculteurs en Europe de l'Ouest. Oxford: Enstone, The Arkleton Trust.

Bryden, J. (1999) *Policymaking for Predominantly Rural Regions: Concept and Issues*, OECD Working Party on Territorial Policy in Rural Areas, Paris.

Bryden, J. and Hart, K. (2001) *Dynamics of Rural Areas (DORA). The International Comparison.* Aberdeen: The Arkleton Centre for Rural Development Research.

Bryden, J.M. and Hart, J.K. (2004) *A New Approach to Rural Development in Europe: Germany, Greece, Scotland and Sweden*, Mellen Studies in Geography Volume 9, The Edwin Mellen Press: Lewiston NY.

Bryden, J., Bell, C., Gilliat, J., Hawkins, E. and MacKinnon, N. (1993) *Farm Household Adjustment in Western Europe 1987–1991*, Final Report on the Research Programme on Farm Structures and Pluriactivity for the Commission of the European Communities, Brussels.

Buckwell, A. (1996) *Moving the CAP towards a more integrated rural policy*, paper to Conference of Agricultural Economics Society of Ireland.

Buckwell, A., Davidova, S., Courboin, V. and Kwiecinski, A. (1995) *Feasibility of an Agricultural Strategy to prepare the countries of Central and Eastern Europe for EU Accession*. Report to European Commission DG 1 Brussels.

Buckwell, A., Blom, J., Commins, P., Hervieu, B., Hofreither, M., von Meyer, H., Rabinowicz, E., Sotte, F. and Sumpsi, J.M. (1997) *Towards a Common Agricultural and Rural Policy for Europe*. Report of an Expert Group, European Commission, Brussels.

Buller, H. (2003) Changing needs, opportunities and threats – the challenges to EU funding of land use and rural development policies, The background to reform, prepared for the LUPG Conference on "Future Policies fir Rural Europe – 2006 and beyond", Brussels, March.

Burrell, A. and Oskam, A. (2000) *Agricultural Policy and Enlargement of the European Union*. Wageningen Agricultural University: Wageningen.

Caraveli, Helen (1997) Environmental Implications of Various Regimes: The Case of Greek Agriculture. *CAP Reform: The Southern Products* (ed. M. Tracy). Belgium: Agricultural Policy Studies.

Cardwell, M. (2004) *The European Model of Agriculture*, Oxford University Press.

Carpathian Ecoregion Initiative, CEI (2001) *The Status of the Carpathians*. A report developed as a part of The Carpathian Ecoregion Initiative.

Case studies of ESPON project 2.1.3 from Austria, Sweden, Greece, Hungary, Slovenia, Scotland

Caskie, P., Davis, J., Campbell, D. and Wallace, M. (2002) *An Economic Study of Farmer Early Retirement and New Entrant Schemes for Northern Ireland*. Belfast: Department of Agricultural and Food Economics, Queen's University Belfast

Caskie, P., Davis, J., Campbell, D. and Wallace, M. (2003) *The EU farmer early retirement scheme: a tool for structural adjustment?* Paper presented at The Agricultural Economics Society, 77th Annual Conference, 11–14 April 2003, University of Plymouth.

Cawley, M., Gillmor, D., Leavy, A. and McDonagh, P. (1995) *Farm Diversification Studies Relating to the West of Ireland*. Dublin: Teagasc.

CEC (Commission of the European Communities) (1981) *Study of the Regional Impact of the Common Agricultural Policy*, Brussels: Regional Policy Series *no.* 21.

CEC (Commission of the European Communities) (1993) *Support for Farms in Mountain, Hill and Less-Favoured Areas*. Brussels: Green Europe *no.* 2/93.

CEC (Commission of the European Communities) (1997) *Rural Developments*, working document, Brussels.

CEC (Commission of the European Communities) (1998), *Agricultural Situation in the CEECs*: Summary Report, DG Agriculture, Brussels.

CEC (Commission of the European Communities) (1999), Directions Towards Sustainable Agriculture, Communication from the Commission to the Council; the European Parliament; the Economic and Social Committeeand the Committee of the Regions, COM (1999) 22 final, Brussels.

CEC (Commission of the European Communities) (2000) *Social Policy Agenda*, COM (2000) 379.

CEC (Commission of the European Communities) (2001) *Environment 2010: Our Future, Our Choice*. COM (2001) 31.

CEC (Commission of the European Communities) (2002) *Mid-Term Review of the Common Agricultural Policy*. Brussels. COM (2002) 394.

CEC (Commission of the European Communities) (2003) Explanatory Memorandum: a Long-term Policy Perspective for Sustainable Agriculture. Brussels. COM (2003) 23.

Central Statistics Office (1994) *Census of Agriculture 1991*. Dublin: CSO.

Central Statistics Office (2002) *Census of Agriculture 2000*. Cork: CSO.

Chanan, G. (1999) *Local Community Involvement: a Handbook for Good Practice*. Dublin: European Foundation for the Improvement of Living and Working Conditions.

Chaplin, H., Davidova, S. and Gorton, M. (2004) Agricultural adjustment and the diversification of farm households and corporate farms in Central Europe, *Journal of Rural Studies* 20, 61-77.

CJC Consulting (2003) *Review of Area-based Less Favoured Area Payments Across Great Britain,* report for Land Use Policy Group (LUPG), May. www.lupg.org.uk.

Colino, J. and Nogueras, P. (1999) La difícil convergencia de las agriculturas europeas. *España, Economía ante el siglo XXI. Ed Ensayo y Pensamiento* (ed. Garcia Delgado).

Commins, P. (1990) Restructuring agriculture in advanced societies: transformation, crisis and responses. *Rural Restructuring: Global Processes and their Responses* (eds T. Marsden, P. Lowe and S. Whatmore). London: David Fulton.

Commins, P. (1996) Agricultural production and the future of small-scale farming, *Poverty in Rural Ireland: A Political Economy Perspective* (eds C. Curtin, T. Haase and H. Tovey), 87–125. Dublin: Oak Tree Press.

Commins, P. (2003) *Trends in Rural Household Incomes*, paper presented to National Rural Development Forum, 2003

Commins, P. and Keane, M. (1995) *Developing the Rural Economy – Problems, Programmes and Prospects,* part 2 of NESC Report no. 97 on New Approaches to Rural Development. Dublin: NESC.

Commins, P. and McDonagh, P. (2002) Enterprise in Rural Areas: Trends and Issues, *Signposts to Rural Change, Proceedings of Rural Development Conference,* 43–61. Dublin: Teagasc.

Commins, P. and Walsh, J.A. (2004) *Agricultural and Rural Development in Ireland: Household Adjustment Strategies and Trends.* NUI Maynooth: National Institute of Regional and Spatial Analysis.

Commons, J.R. (1931) Institutional Economics. *American Economic Review*, 21, 648–657.

Community Development Foundation and Urban Forum (2001). *The LSP Guide to Local Strategic Partnerships.*

Connolly, L., Burke, T. and Roche, M. (2001) *National Farm Survey*. Dublin: Teagasc.

Connolly, L. (2002) Costs and Margins in Organic Production in Comparison with Conventional Production, *Signposts to Rural Change, Proceedings of Rural Development Conference,* 91–98. Dublin: Teagasc.

Copus, A.K. (2001) From core-periphery to polycentric development: concepts of spatial and aspatial peripherality. *European Planning Studies* 9(4), 539–552.

Copus A.K. (ed.) (2004) *Aspatial Peripherality, Innovation and the Rural Economy,* Final Report of EU Fifth Framework project QLK-2000-00783 (AsPIRE) downloadable from http://www.sac.ac.uk/ aspire.

COR (Committee of the Regions) (1999) *Opinion on the European Spatial Development Perspective*, CdR 266/98 fin, Brussels.

Council of the EU (2003) *Multiannual Strategic Programme.* Document POLGEN 85, Brussels, 8 December. http://www.eu2004.nl/default.asp?CMS_TCP=tcpAsset&id=1CE82DB6F32A4717B651 4E3646682880

Courades, M. (2004) *Leader programme – lessons for new rural Europe,* paper to EU conference on Improving Living Conditions and Quality of Life in Rural Europe, Westport, Ireland, 1 June 2004.

CPMR (1997) Towards a Balanced Europe: From Peripheries to Large Integrated Maritime Units, Rennes.

Crowley, C. and Walsh, J. (2003) County level adjustments in agricultural production in the context of CAP reforms in Ireland 1991–2000. Department of Geography, NUI Maynooth.

Crowley, C., Meredith, D. and Walsh, J. (2004) Population and agricultural change in rural Ireland 1991-2002, *Rural Futures, Proceedings of Rural Development Conference,* (ed. E. Pitts), 17–34. Dublin: Teagasc.

Cunder, T. (2004) Less-Favoured Areas Scheme In Slovenia, Case Study ESPON Project 2.1.3. Ljubljana.

CURS (Centre for Urban and Regional Studies) (2004) *Urban-Rural Relations in Europe*, ESPON Project 1.1.2 Final Report. Helsinki University of Technology.

Dalton, G.E. (2004) *The Special Pre-accession Programme for Agricultural and Rural Development in Poland (SAPARD)*, case study paper for ESPON TPG 2.1.3, University of Aberdeen, UK. http://www.abdn.ac.uk/arkleton/publications/index.shtml

Daucé, P., Leturqu, F. and Quinqu, M. (1999) Impact of the second early retirement scheme on young farmers' setting-up, *Economie rurale* 253, 51–57. (in French).

Davoudi, S. (2002) Polycentricity: – modelling or determining reality? *Town and Country Planning Journal* 71(4), 114–117.

Davoudi, S. (2003) Polycentricity in European spatial planning: from an analytical tool to a normative agenda. *European Planning Studies,* 11(8), 979–999.

Davoudi, S. and Stead, D. (2002) Urban-rural relationships: an introduction and brief history. *Built Environment* 28(4), 269–277.

Dax, T. (1995) Strukturelle Veränderungen im ländlichen Raum Europas. *Strukturen in Landwirtschaft und Agribusiness* (eds W. Schneeberger and H.K. Wytrzens), 37–48. Dokumentation der 4. Wien: ÖGA-Jahrestagung.

Dax, T. (1999) Entwicklung des landlichen Raumes – Kompromiss oder 'Zweiter Pfeiler der Gemeinsamen Agrarpolitik'. *Landwirtschaft 99, der kritische Agrarbericht* (ed. E.V. Agrarbundnis). Kassel, 44–50.

Dax, T. (2001) *The quest for countryside support schemes for mountain areas in Central and Eastern European Countries.* Paper to the 41st Congress of the European Regional Science Association, ERSA, Zagreb.

Dax, T. (2002a) *Rural development policy from an EU perspective*, paper at the XXI Summer Course – XIV European Courses "Desarollo rural y gestión territorial", University of the Basque Country, Donostia – San Sebastián, 1–2 August 2002.

Dax, T. (2002b) The impact of the policy framework on dynamics of mountain farming in Austria, paper to the Fifth IFSA European Symposium, International Farming Systems Association, *Farming and Rural Systems Research and Extension, Local Identities and Globalisation* pre-proceedings, 8–11 April (ed. IFSA European Group), 229–238, Firenze.

Dax, T. (2004) *The Impact Of EU Policies On Mountain Development In Austria*, paper at the Regional Studies Association – International Conference "Europe at the Margins: EU Regional Policy, Peripherality and Rurality", 15–16 April 2004, Angers, France http://www.regional-studies-assoc.ac.uk/events/presentations04/dax.pdf

Dax, T. and Hellegers, P. (2000) Policies for less-favoured areas. *CAP Regimes and the European Countryside, Prospects for Integration between Agricultural, Regional and Environmental Policies* (ed. F. Brouwer and P. Lowe), 179–197. Wallingford: CAB International. http://www.cabi-publishing.org/Bookshop/ReadingRoom/ 0851993540/3540Ch11.pdf

Dax, T. and Herbertshuber, M. (2002) Regional and rural development in Austria and its influence on leadership and local power. *Leadership and Local Power in Rural Development* (ed. K. Halfacree, I. Kovach and R. Woodward), 203–229. Aldershot, Hampshire: Ashgate.

Dax, T. and Hovorka, G. (2002) Innovative Structures for the Sustainable Development of Mountainous Areas. The Austrian Case: Almenland and Teichalm-Sommeralm (Alpine Pasture Area), Discussion Paper 5 of the ISDEMA Project, Vienna.

Dax, T. and Hovorka, G. (2003) Integrated development for multifunctional landscapes in mountain areas, *Sustaining Agriculture and the Rural Economy: Governance, Policy and Multifunctionality* (ed. F. Brouwer), Camberley, UK: Edward Elgar Publishing.

Dax, T., Loibl, E. and Oedl-Wieser, T. (eds) (1995) *Pluriactivity and Rural Development, Theoretical Framework*, research report *no.* 34 of Bundesanstalt für Bergbauernfragen, Wien.

Dax, T., Hovorka, G. and Wiesinger, G. (2002) Perspektiven für die Politik zur Entwicklung des ländlichen Raums. *Der GAP-Reformbedarf aus österreichischer Perspektive* (ed. M. Hofreither). Wien: Forschungsbericht im Auftrag des BMLFUW.

de Boe P. and Hanquet, F. (2004) *Towards an ESPON Glossary*, ESPON Project 3.1 Final Report Part C - Annex, Brussels. www.espon.lu.

Department of Agriculture and Food (1999) *Evaluation of the Rural Environmental Protection Scheme.* Dublin: Government Publications.

Dethier, J. L., Saraceno, E., Bontron, J.C. and von Meyer, H. (1999) *Ex-Post Evaluation of the LEADER I Community Initiative 1989–1993*, General Report, Brussels.

DG Agriculture (2003) *Overview of the Implementation of Rural Development Policy 2000–2006* http://europa.eu.int/comm/agriculture/publi/fact/rurdev2003/ov_en.pdf

DG Regio (2004) Interim Territorial Cohesion Report: preliminary results of ESPON and EU Commission studies.

Dortmund (2003) Unpublished project reports from the European research project "newrur" ("urbaN prEssure on RURal areas")

Dwyer, J., Baldock, D., Beaufoy, G., Bennett, H., Lowe, P. and Ward, N. (2002) *Europe's Rural Futures – the Nature of Rural Development: rural development in an enlarging Europe.* Land Use Policy Group of Great Britain and WWF Europe with the Institute for European Environmental Policy, London, 52–54.

Dwyer, J., Baldock, D., Beaufoy, G., Bennett, H., Lowe, P. and Ward, N. (2003) *Europe's Rural Future – The Nature of Rural Development II: Rural Development in an Enlarging European Union.* London: Land Use Policy Group and WWF Europe, Institute for European Environmental Policy, London.

ECA (European Court of Auditors) (2003) *Rural Development: Support For Less-Favoured Areas.* Special Report no. 4/2003 with Commission's Replies. *Official Journal of the European Union* 46 C 151/01, 27 June. Luxembourg. Available at http://www.eca.eu.int/EN/reports_opinions.htm

Eckstein, K., Hoffmann, H. and Gloeckler, J. (2004) *Mid Term Evaluation of the Bavarian Agri-Environmental Programme,* paper at the 87th EAAE-Seminar: Assessing rural development policies of the CAP, 21–23 April, Wien.

Edmond, H. and Crabtree, JR. (1994) Regional variation in Scottish pluriactivity: the socio-economic context for different types of non-farming activity. *Scottish Geographical Magazine,* 110(2), 76–84.

Edmond, H., Corcoran, K. and Crabtree, J.R. (1993) Modelling Locational Access to Markets for Pluriactivity: a Study in the Grampian Region of Scotland. *Journal of Rural Studies,* 9(4), 339–349.

EEA (European Environment Agency) (2001) Towards Agri-Environmental Indicators: Integrating Statistical and Administrative Data with Land Cover Information, Topic Report no. 6, Copenhagen.

EEA (European Environment Agency) (2004) Agriculture and the Environment in the EU Accession Countries – Implications of Applying the EU, Environmental Issue Report no. 37, Copenhagen.

Efstratoglou, S., Psaltopoulos, D., Thomson, K.J., Snowdon, P., Daouli, J., Skuras, D., Kola, J. and Nokkala, M. (1998) *Structural Policy Effects in Remote Rural Areas Lagging behind in Development,* final report to European Commission on STREFF project FAIR3 CT96-1554, Ch. 6.

Elbe, S., Kroes, G., Krott, M., Böcher, M., Hubo, C. and Schubert, D. (2003) Berücksichtigung von Naturschutzzielen in der Programmbewertung und -entwicklung (BNPro). Ergebnisse der Breitenanalyse. Zwischenbericht zum FuE-Vorhaben FKZ 802 82 050. 203 Seiten. Darmstadt. (unpublished).

Emerson, H. and Gillmor, D. (1999) The Rural Environment Protection Scheme of the Republic of Ireland, *Land Use Policy,* 16(4), 235–245

Errington, A. and Lobley, M. (2002) *Handing Over the Reins: A Comparative Study of Intergenerational Farm Transfers in England, France, Canada and the USA.* Conference Paper at the Agricultural Economics Society, Aberystwyth, 8–11 April 2002.

Esparcia, J. and Noguera, J. (2004) PRODER Andalucia, Monograph Spain, in: ÖIR (coord.), Methods for and Success of Mainstreaming Leader Innovations and Approach into Rural Development Programmes, commissioned by European Commission, DG Agriculture, Unit G4, Wien, pp.I-19-I-35.

European Commission (1994) *The Economics of the Common Agricultural Policy (CAP),* Reports and Studies no. 5. Luxembourg.

European Commission (1998) *Agricultural Situation and Prospects in the Central and Eastern European Countries.* Summary report by the Directorate General for Agriculture (DG VI). Brussels: Office for Official Publications of the European Communities.

European Commission (1999) ESDP, European Spatial Development Perspective: Towards Balanced and Sustainable Development of the Territory of the EU. Luxembourg: Office for Official Publications of the European Communities.

European Commission (2001a) Sustainable Agriculture in Central and Eastern European Countries (CEESA) – Description of Current Farming Systems in Central and Eastern European Countries. Project under EU 5th framework programme. Budapest.

European Commission (2001b) SAPARD Annual Report – Year 2000. Report from the Commission to the European Parliament, the Council, the Economic and Social Committee and the Committee of the Regions. COM (2001) 341 final. Brussels.

European Commission (2001c) *Unity, Solidarity, diversity for Europe, its people and its territory: Second report on Economic and Social Cohesion*. Luxembourg: Office for Official Publications of the European Communities, 2001. ISBN 92-894-0572-4.

European Commission (2002a) Enlargement and Agriculture: Successfully Integrating the New Member States into the CAP. SEC (2002) 95 final. Brussels.

European Commission (2002b) *Mid-Term Review of the CAP*, COM (2002) 394 Final. Brussels.

European Commission (2002c) *SAPARD Annual Report – Year 2001*. COM (2002) 434 final. Brussels.

European Commission (2002d) Enlargement: Successfully Integrating the New Member States into the CAP, Newsletter *no.* 42, March.

European Commission (2003b) *Rural Development in the European Union*. Fact Sheet, Luxembourg.

European Commission (2004a) *Rural Development in the EU*, Memo 04/180, 15 July, Brussels.

European Commission (2004b) *A New Partnership for Cohesion – Convergence, Competitiveness, Cooperation*. Third Report on Economic and Social Cohesion, Brussels.

European Commission (2004c) *New Perspectives for Rural Development*. Luxembourg: Office for Official Publications of the European Communities. http://europa.eu.int/comm/agriculture/publi/fact/rurdev/refprop_en.pdf

European Commission, Directorate General for Agriculture (1996) *The Cork Declaration - a living countryside*, Brussels. http://europa.eu.int/comm/agriculture/rur/cork_en.htm

European Commission, Directorate General for Agriculture (1998) *Evaluation of Agri-Environment Programmes CAP 2000*, Working Document, VI/7655/98 Brussels

European Commission, Directorate General for Agriculture (2000a) EU and Enlargement – Pre-Accession Instruments: Focus on Agriculture, Brussels.

European Commission, Directorate General for Agriculture (2000b) Indicators for the Integration of Environment Concerns into the Common Agricultural Policy, Brussels.

European Commission, Directorate General for Agriculture (2002) Analysis of the Impact on Agricultural Markets and Incomes of EU Enlargement to the CEECs. Brussels.

European Commission, Directorate General for Agriculture (2003a) *Key Developments in the Agri-Food Chain and on Restructuring and Privatisation in the CEE Candidate Countries* February. http://europa.eu.int/comm/agriculture/publi/reports/agrifoodchain/2002_en.pdf

European Commission, Directorate General for Agriculture (2003b) *Reform of the Common Agricultural Policy: Long-Term Perspective for Sustainable Agriculture*: Impact Analysis. http://europa.eu.int/comm/agriculture/publi/reports/reformimpact/rep_en.pdf

European Commission, Directorate-General for Agriculture (2003c) *Mid-Term Review of the Common Agricultural Policy: July 2002 Proposals, Impact Analyses*, February. Brussels. http://europa.eu.int/comm/agriculture/publi/reports/mtrimpact/rep_en.pdf.

Evans, N., Morris, C. and Winter, M. (2002) Conceptualizing agriculture: a critique of post-productivisim as the new orthodoxy. *Progress in Human Geography* 26(3), 313-32.

Faludi A. and Waterhout, B. (2002) The Making of the European Spatial Development Perspective, Routledge: London.

Feehan, J., Flynn, M., Carton, O., Culleton, N., Gillmor, D. and Kavanagh, B. (2002) REPS and the UN Convention on Biodiversity: the importance of agri-environment to biodiversity conservation in Ireland. *Achievement and Challenge: Rio + 10 and Ireland,* (eds F. Convery and J. Feehan), 29–36. Dublin: The Environmental Institute, University College.

Fennell, R. (1981) Farm Succession in the European Community, *Sociologia Ruralis* 21 (1), 19–42.

Fennell, R. (1997) The Common Agricultural Policy: Continuity and Change, Oxford: Clarendon Press

Ferenczi, T. (2003) *Rural Development in Hungary*. Budapest: Universidad de Budapest de cc Economicas y Administracion Publica.

Frawley, J. (1998) The Impact of Direct Payments at Farm Level: a County Study. Dublin: Teagasc.

Frawley, J. and Phelan, G. (2002) Changing Agriculture: Impacts on Rural Development, *Signposts to Rural Change: Proceedings of Rural Development Conference,* 20–42. Dublin: Teagasc.

Frawley, J., Commins, P., Scott, S. and Trace, F. (2000). *Low Income Farm Households: Incidence, Characteristics and Policies*. Dublin: Oak Tree Press.

Gallardo Cobos, R. (2001) Análisis de los efectos de las reformas de la política agraria común y de la viabilidad de las estrategias adaptativas en sistemas agrarios del Valle del Guadalquivir. Escuela Técnica Superior de Ingenieros Agrónomos y de Montes. PhD Dissertation. Universidad de Córdoba.

Gallardo Cobos, R., Ramos Real, F. and y Ramos Real, F. (2002) Perturbaciones provocadas por la nueva PAC en las decisiones de ajuste estratégico en sistemas agrarios andaluces. *Economía Agraria y Recursos Naturales*, 2(1), 131–151

Gasson, R. and Errington, A. (1993) *The Farm Family Business*. Wallingford, Oxon: CAB International.

Gatzweiler, F.W. (2003) *Institutional Change in Central and Eastern European Agriculture and Environment*, Synopsis of the Central and Eastern European Sustainable Agriculture Project (CEESA), CEESA/FAO Series, Volume 4, Roma and Berlin. http://www.fao.org/regional/ SEUR/CEESA_Vol4_en.pdf

Gatzweiler, S. and Bäckmann, Z. (2001) Analysing institutions, policies & farming systems for sustainable agriculture in central and eastern European countries in transition. CEESA Discussion Paper *no.* 2/5/2001.

Gerhardter, G. and Gruber, M. (2001), Regionalförderung als Lernprozess, Evaluierung der Förderungen des Bundeskanzleramtes für eigenständige Regionalentwicklung, Schriften zur Regionalpolitik und Raumordnung *no.* 32, Wien

Gillmor, D.A. (1999) The scheme of early retirement from farming in the Republic of Ireland, *Irish Geography,* 32(2), 76–86

Gindl, M., Stuppäck, S. and Wukovitsch, F. (2001) *"Good Practice" Partnership for Sustainable Urban Tourism. "Cheese Route Bregenzerwald", Vorarlberg, Austria SUT Governance*. Research project of Key Action 4 within the Fifth Framework Program, Vienna.

Goetz, S.J. and Debertin, D.L. (1996) Rural population decline in the 1980s: Impacts of farm structure and federal farm programs, *American Journal Of Agricultural Economics*, 78 (3), 517–529.

Groier, M. (2004) Regional effects of the Austrian Agri-Environmental Programme ÖOPUL in the case study area Wien: Bludenz-Brehgenzerw Wald, BA füurfur Bergbauernfragen.

Groier, M. and Hofer, O. (2002) *Evaluierung des ÖPUL 95 und ÖPUL 98 Bericht 2001*, Wien: Beirat für die Evaluierung des Umweltprogramms.

Hair, J.F., Anderson, R.E., Tatham, R.L. and Black, W.C. (1995) *Multivariate Data Analysis with Readings*, 4th edition, Upper Saddle River, NJ: Prentice Hall.

Hajer, M. (1995) The Politics of Environmental Discourse: Ecological Modernisation and the Policy Process. Oxford: Clarendon Press.

Halfacree K.H. (1999) A new space or spatial effacement? Alternative futures for the post-productivist countryside. *Reshaping the Countryside: Perceptions and Processes of Rural Change*. (eds N.S. Walford, J. Everitt and D. Napton), 67–76. Wallingford: CAB International.

Halfacree, K., Kovach, I. and Woodard, R. (eds) (2002) *Leadership and Local Power in European Rural Development*, Perspectives on Rural Policy and Planning, Aldershot: Ashgate.

Hantrais, L. (2000) *Social Policy in the European Union*. London: Macmillan Press.

Harte, L.N. and O'Connell, J.J. (2002) Agri-environmental Payments: Income supports with Cross Compliance or Payments for Environmental Outputs? *Achievement and Challenge: Rio + 10 and Ireland* (eds F. Convery and J. Feehan), 40–47. Dublin: The Environmental Institute, University College Dublin.

Hausmann, C. (ed.) (1996) Lo sviluppo rurale, Turismo rurale, agriturismo, prodotti agroalimentari, LEADER II quaderno informative n.4, Roma.

Healey, P. (2004) The Treatment of Space and Place in the New Strategic Spatial Planning in Europe, *International Journal of Urban and Regional Research*, 28(1), 45–67.

Hesina, W., Knoflacher, M., Wagner, P., Lechner, F., Bergmann, N., Pfefferkorn, W. and Resch, A. (2002) *Ex-post-Evaluierung der Ziel 5b- und LEADER II-Programme 1995–1999 in Österreich,* ÖROK schriftenreihe Nr. 161, Wien. (2 volumes).

Hoggart, K., Black, R. and Buller, H. (1995) *Rural Europe: Identity and Change*. London: Edward Arnold.

Hovorka, G. (2003) Evaluierung des Österreichischen Programms für die Entwicklung des ländlichen Raums: Teilbereich Förderung landwirtschaftlich benachteiligter Gebiete (Kapitel V), Wien.

Hovorka, G. (2004) Evaluation of the Compensatory Allowances Scheme under the EU Regulation 1257/99 in Austria and in Other EU Member States, paper at 87th EAAE conference, 21–23 April 2004, Vienna.

Howe, K., McInerney, J., Traill Thomson, J., Turner, M. and Whitaker, J. (1996) *Resource Use Adjustment in the Rural Economy*. Report no. 244. University of Exeter, Agricultural Economics Unit.

Hunter, B., Commins, P. and McDonagh, P. (2004) An approach to developing tourism in rural areas: lessons learned from a regional case study, *Rural Futures, Proceedings of Rural Development Conference*, (ed. E. Pitts), 53–70. Dublin: Teagasc.

Hyyryläinen, T. and Pylkkänen, P. (2004) POMO and POMO+, Monograph Finland, in: ÖIR (coord.), Methods for and Success of Mainstreaming Leader Innovations and Approach into Rural Development Programmes, commissioned by European Commission, DG Agriculture, Unit G4, Wien, I-37–I-50.

IAMO (2004) *The Future of Rural Areas in the CEE New Member States*, Report by Network of Independent Experts in the CEE Candidate Countries, European Commission, 20 April 2004.

IDARA (2002) Identification of the Critical Socio-Economic Problems facing rural CEEC – policy proposals. Galway: IDARA.

IDARA (2003) Strategies for integrated development of agriculture and rural areas in Central European Countries: the integrated IDARA results on the CAP and rural policies, draft version 10/07/03, Fifth Framework Programme, QLRT-1526. http://www.agp.uni-bonn.de/agpo/rsrch/idara/D17_Integrated Results.doc

Ilbery, B.W. (1981) *Western Europe: A Systematic Human Geography*. Oxford: Oxford University Press.

INEA (2002a) Osservatorio sulle Politiche Agricole dell'UE, *Le politiche agricole dell'Unione Europea, Rapporto 2001–02*, Osservatorio sulle Politiche Agricole dell'UE, Settembre, http://www.inea.it/pdf/pac2002.pdf

INEA (2002b) The Mid-Term Review of the Common Agricultural Policy: Assessing the Effects of the Commission Proposals. Rome: INEA.

Ingersent, K. and Rayner, A. J. (1999) *Agricultural Policy in Western Europe and the United States*. Edward Elgar: Cheltenham.

Isermeyer, F., Kleinhanß, W., Manegold, D., Mehl, P., Nieberg, H., Offermann, F., Osterburg, B., Schrader, H., and Seifert, K. (1999) Auswirkungen der Beschlusse zur Agenda 2000 auf die deutsche Land- und Forstwirtschaft : Antworten auf den Fragenkatalog anläßlich der öffentlichen Anhörung des Ernährungsausschusses des Deutschen Bundestages am 16.6.1999 [online]. Braunschweig : FAL. http://www.bal.fal.de/download/ea_gesamt.pdf (Volltext)

IRDSP (Institute for Regional Development and Structural Planning) (2004) *Territorial Effects of the "Acquis Communitaire", Pre-Accession Aid and Phare/Tacis/Meda Programmes*, Third Interim Report, ESPON Project 2.2.2, Erkner (Germany). www.espon.lu.

ISTAT (1998), Le aziende agroturistiche in Italia, Roma.

Johansson, B. and Quigley, J.M. (2004) Agglomeration and Networks in Spatial Economics. *Papers in Regional Science* 83, 165–176.

Kearney Associates, Fitzpatrick Associates and Walsh, J. (2000) *Evaluation of LEADER 11 Programme in Ireland*, report to Department of Agriculture and Food, Dublin.

Kearney, B., Boyle, G. and Walsh, J. (1995a) *EU LEADER 1 Initiative in Ireland: Evaluation and Recommendations*, Dublin: Department of Agriculture, Food and Forestry.

Kearney, B., Boyle. G. and Walsh, J. (1995b) *Evaluation of Impacts of Compensatory Allowances in Less Favoured Areas of Ireland*. Dublin: Report to Department of Agriculture, Food and Forestry.

Kilkenny, M. (2004) *Geography, Agriculture and Rural Development*, plenary address to 78th Agricultural Economics Society Annual Conference, Imperial College London

Koutsomiti, E. (2000) Structural Measures of Common Agricultural Policy and Sustainable Development of Insular Districts: Early Retirement, New Farmers and Environment in the Island of Lesvos, University of Aegean, MSc Dissertation (in Greek)

Koutsouris, A. (ed.) (2003) *Innovative Structures for the Sustainable Development of Mountainous Areas*, Proceedings of the ISDEMA Conference in Thessaloniki, Greece, Volume II, 8–9 November 2002, National and Kapodistrian University of Athens.

Lafferty, S., Commins, P. and Walsh, J. (1999) Irish Agriculture in Transition – A Census Atlas of Agriculture in the Republic of Ireland. Dublin: Teagasc and NUI Maynooth.

Lamaison, P. (1988) La diversité des modes de transmission: une géographie tencace. *Etudes Rurales* (110/111/112), 119–175.

Loesch, R. and Meimberg, R. (1986) *Der alternative Landbau in der Bundesrepublik Deutschland - Abgrenzung, Produktion, Vermarktung* - Ifo-Institut fuer Wirtschaftsforschung e.V. Muenchen 1986

Louloudis, L., Maraveyas, N. and Martinos, N. (1993) The social dimension of CAP. *Contemporary Dimensions of Social Policy*. Athens: Sakis Karagiorgas Institution (in Greek)

Lowe, P. and Brouwer, F. (2000) Agenda 2000: a Wasted Opportunity? *CAP Regimes and the European Countryside, Prospects for Integration between Agricultural, Regional and Environmental Policies* (eds F. Brouwer and P. Lowe), 321–334. Wallingford, Oxon: CABI Publishing.

Lowe, P., Ray, C., Ward, N., Wood, D. and Woodward, R. (1999) *Participation in Rural Development*, European Foundation for the Improvement of Living and Working Conditions, Dublin. Luxembourg: Office for Official Publications of the European Community.

Lowe P., Buller H. and Ward N. (2002) Setting the Next Agenda? British and French Approaches to the Second Pillar of the CAP. *Journal of Rural Studies*, 18, 1–17.

Machold, I. (2004) LEADER – good practice in territorial rural development, synthesis of case studies. Wien: Bundesanstalt fuer Bergbauernfragen.

Maitz, J. (2004) LEADER als Modell für Ungarn, *LEADER Magazin Österreich* 1_04, Wien, 38–39.

Mantino, F. (2003) *The Second Pillar: Allocation of Resources, Programming and Management of Rural Development Policy*, prepared for the Land Use Policy Group Conference on "Future Policies for Rural Europe – 2006 and beyond", Brussels.

Marsden, T. (2003) *The Condition of Rural Sustainability*, Assen: Van Gorcum.

Matthews, A. (2002) Has agricultural policy responded to the Rio challenge? *Achievement and Challenge: Rio + 10 and Ireland* (eds F. Convery and J. Feehan), 73–82. Dublin: The Environmental Institute, University College Dublin.

McCann, P. and Shefer, D. (2004) Location, Agglomeration and Infrastructure, *Papers in Regional Science* 83, 177–196.

McDonagh, P. and Commins, P. (1999) Globalization and rural development: demographic revitalisation, entrepreneurs and small business formation in the west of Ireland, *Local Responses to Global Integration* (eds C. Kasimis and A. G. Papadapoulos), 179–202. Aldershot: Ashgate.

McDonagh, P., Commins, P. and Gillmor, D. (1999) Supporting Entrepreneurship in Marginal Rural Areas, *Local Enterprise on the North Atlantic Margin*, (eds R. Byron and J. Hutson), 15–44. Aldershot: Ashgate.

McHugh, C. and Walsh, J. (2000) A multivariate typology of rural areas in Ireland, *Irish Rural Structure and Gaeltacht Areas* (Fitzpatrick Associates report for National Spatial Strategy), 5–33. Dublin: Department of Environment and Local Government.

McNally, S. (2001) Farm diversification in England and Wales – what can we learn from the farm business survey? *Journal of Rural Studies* 17, 247–257.

Ministry of Agriculture (1994) Directorate of Physical Planning and Environmental Protection. General Framework for Application of Reg. 2078/92. Athens, Greece (in Greek).

Ministry of Agriculture (2001) Performance Audit Report, Information on Farmers' Retirement Support in Budget Bills (Dno: 205/54/01), Finland. http://www.vtv.fi/vtv/julkaisu_.en.nsf

Ministry of Agriculture and Rural Development (2004) *Agriculture and Rural Development Operational Programme, Budapest, 2004–2006* Republic of Hungary, Budapest.

Mora, R., and San Juan, C. (2003) Geographical specialization in Spanish agriculture before and after integration in the European Union. Regional Science and Urban Economics.

Moseley, M. (ed.) (2003) Local Partnerships for Rural Development: the European Experience. Wallingford, Oxford: CABI Publishing.

Murphy, S. (1997) Evaluation of the Scheme of Installation Aid for Young Farmers: Including an Assessment of the Effects of the Scheme of Early Retirement from Farming. Dublin, Ireland: Analysis and Evaluation Unit, Department of Agriculture, Food and Rural Development.

Naylor, E.L. (1982) Retirement policy in French agriculture. *Journal of Agricultural Economics* 33(1), 25–36.

Newby, H. (2004) *An Insider's View from the Outside: Rural Sociology and the future of Rural Society.* Paper to XI World Congress of Rural Sociology, Trondheim, Norway.

Nordregio (2003a) *Analysis of Mountain Areas in the European Union and in the Applicant Countries,* Second interim report, European Commission DG Regio, Stockholm, 28th February.

Nordregio (2003b) The Role, Specific Situation and Potentials of Urban Areas as Nodes of Polycentric Development, ESPON project 1.1.1, executive summary, July.

Nordregio (2004) Mountain Areas in Europe, Analysis of Mountain Areas in EU Member States, Acceding and Other European Countries, Nordregio Report 2004:1, Stockholm.

Norman, G.R. and Streiner, D.L. (1999) *PDQ Statistics,* 2nd edition, Hamilton-London-Saint Louis: B.C Decker Inc.

O'Cinneide, M., Keane, M. and Wiium, V. (1999) Optimising the contribution of private forestry in the sustainable development of rural Ireland, *Local Enterprise on the North Atlantic Margin,* (eds R. Byron and J. Hutson), 205–223. Aldershot: Ashgate.

OECD (1993a) Territorial Development and Structural Change: A New Perspective on Adjustment and Reform. Paris: OECD.

OECD (1993b) What Future For Our Countryside? A Rural Development Policy. Paris: OECD.

OECD (1996a) Territorial Indicators of Employment, Focusing on Rural Development. Paris: OECD.

OECD (1996b) Better Policies for Rural Development. Paris: OECD

OECD (1997) Agricultural Policies in Transition Economics. Monitoring and Evaluation 1997. Paris: OECD.

OECD (1998a) Agricultural Policy Reform and the Rural Economy in OECD Countries. Paris: OECD.

OECD (1998b) Best Practices in Local Development. Paris: OECD.

OECD (1998c) Rural Amenity In Austria, a Case Study Of Cultural Landscape. Paris: OECD.

OECD (1999) Cultivating Rural Amenities. An Economic Development Perspective. Paris: OECD.

OECD (2002a) *Factors of Growth in Predominantly Rural Regions,* Working Party on Territorial Policy in Rural Areas, document DT/TDPC/RUR(2002)3, Paris.

OECD (2002b) *Identifying the Determinants of Regional Performances,* Working Party on Territorial Indicators, Paris.

OECD (2002c) Siena, Italy, *OECD Territorial Reviews,* Paris.

OECD (2002d) Methodology for the Measurement of Support and Use in Policy Evaluation – 2002, Paris.

OECD (2003a) Agricultural Policies in OECD Countries. Paris: OECD.

OECD (2003b) *Agricultural and Rural Development: Policies in the Baltic Countries,* Paris: Organisation for Economic Cooperation and Development

OECD (2003c) Farm Household Income, Issues and Policy Responses. Paris: OECD.

OECD (2004) Agricultural Policies at a Glance. Paris: OECD.

ÖIR (Österreichisches Institut für Raumplanung) (2003) *Ex-post Evaluation of the Community Initiative LEADER II.* Final Report to the European Commission DG Agriculture, Vienna.

ÖIR (Österreichisches Institut für Raumplanung) (2004) *Methods For and Success of Mainstreaming Leader Innovations and Approach into Rural Development Programmes.* Report to the European Commission, DG Agriculture, Vienna.

ÖROK (Österreichische Raumordnungskonferenz) (2002) Ex-post-Evaluierung der Ziel 5b- und LEADER II-Programme 1995 – 1999 in Österreich, Band II – LEADER II, Schriftenreihe Nr. 161/II, Wien.

Paniagua Mazorra, A. (2000) Analysis of the evolution of farmers' early retirement policy in Spain: The case of Castille and León, *Land Use Policy* 17, 113–120.

Paniagua Mazorra, A. (2001) Agri-environmental policy in Spain. The agenda of socio-political developments at the national, regional and local levels, *Journal of Rural Studies* 17, 81–97.

Parr, J.B. (2003) *Reinventing regions? The case of the polycentric urban regions.* Paper at the Regional Studies Association conference, Pisa, Italy, 12–15 April 2003.

Pérez Yruela, M., Sumpsi, J.M., Bardají, I. and Guerrero, G. (2000) *La Nueva Concepción del Desarrollo Rural: Estudio de Casos.* Colección Politeya. Estudios de Política y Sociedad. Consejo Superior de Investigaciones Científicas.

Perrier-Cornet, P., Blanc, M., Cavailhes, J., Dauce, P. and Le Hey, A. (1991) La Transmission Des Exploitations Agricoles et L'Installation des Agriculteurs dans la CEE (Farm Take-Over and Farm Entrance within the E.E.C). Monographies Nationales. Brussels: CCE DG VI, and Dijon: INRA ESR.

Peters, R. (2002) Shaping the Second Pillar of the CAP. *EuroChoices* 1(2), 20–21.

Pettit, A. (2001) *The role of small and medium-sized towns in rural development*, FP5 project (2001–2004) LIFE QUALITY, coordination University of Reading.

Pezzini, M. (2003) *Governance In Rural Areas Under Restructuring*. OECD Working Party on Territorial Policy in Rural Areas, Paris.

Philip, L. and Shucksmith, M. (2003) Conceptualizing Social Exclusion in Rural Britain, *European Planning Studies*, 11(4), 461–480.

Pietola, K., Väre, M. and Lansink, A.O. (2003) Timing and type of exit from farming: farmers' early retirement programmes in Finland, *European Review of Agricultural Economics* 30(1), 99–116.

Potter, C. and Lobley, M. (1992) Ageing and succession on family farms: the impact on decision-making and land use. *Sociologia Ruralis* 32(2/3), 317–334.

Potter, C. and Lobley, M. (1996) The farm family life cycle, succession paths and environmental change in Britain's countryside. *Journal of Agricultural Economics* 47(2), 172–190.

Pouliquen, A. (2001) Competitiveness and farm incomes in the CEEC agri-food sectors – Implications before and after accession for EU markets and policies. Report prepared for DG AGRI, October 2001.

Pouw, X. (2000) L'impact Environmental de la Culture du Mais dans l'Union Europenne

Prazan, J., Ratinger, T., Krumalova, V., Lowe, P. and Zelei, A. (2003) Maintaining High Nature Value Landscapes in an Enlarge Europe: A Comparative Analysis of the Czech Republic, Hungary and Slovenia, Institutional Change in Central and Eastern European Agriculture and Environment, CEESA/FAO Series, Volume 1, Roma and Berlin. http://www.fao.org/regional/SEUR/CEESA _Vol1_en.pdf

Pretty, J.N., Brett, C., Gee, D., Hine, R.E., Mason, C.F., Morison, J.I.L., Raven, H., Rayment, M.D. and van der Bijl, G. (2000) An assessment of the total external costs of UK agriculture. *Agricultural Systems* 65(2), 113–136.

Privitera, D. (2004) EU-policy for rural development. Some implications for agro-tourism, paper at the 87th EAAE-seminar "Assessing rural development policies of the CAP", 21–23 April, Wien.

Putnam, R. (1993) *Making Democracy Work. Civic Traditions in Modern Italy*, Princeton: Princeton University Press.

Quendler, T. (1997) Grundstückspacht als Faktor des Agrarstrukturwandels, *Grüner Bericht 1996*, 59–60, Wien: BMLFUW.

Rabinowicz, E., Thomson, K. J. and Nalin, E. (2001) *Subsidiarity, the CAP and EU Enlargement*, Swedish Institute for Food and Agricultural Economics: Lund.

Ratinger, T. and Křůmalová V. (2002) Provision of environmental goods on potentially abandoned land – the White Carpathians protected landscape area. CEESA Discussion Paper no. 6/2002.

Rietveld P and Vickerman R. (2004) Transport in regional science: The "death of distance" is premature, *Papers in Regional Science* 83, 229–248.

Ritson, C. and Harvey, D.R. (1997) *The Common Agricultural Policy* (2nd edition). CAB International: Wallingford.

Robert, J., Stumm, T., Vet. J.M., Reincke, G.J., Hollanders, M. and Figueiredo, M.A. (2001) *Spatial Impacts of Community Policies and the Costs of Non-coordination*, report to DG Regio, European Commission, Brussels.

Roberts, D., Phimister, E. and Gilbert, A. (2001) *Pluriactivity and Farm Incomes in Scotland: Longitudinal Analysis using the Farm Accounts Survey*, Final Report for the Scottish Executive Environment and Rural Affairs Department, (MLU/761/00).

Robinson, G. (2004) *Geographies of Agriculture*, Harlow: Pearson.

Rogers, C.S. and Salamon, S. (1983) Inheritance and Social Organisation among Family Farmers. *American Ethnologist* 10(3), 529–550.

Ross Gordon Consultants SPRL, Belgium (2000) *The Future of Young Farmers in the European Union.* European Parliament L-2929 Luxembourg. Directorate General for Research. Agriculture, Forestry and Rural Development Series. AGRI 134 EN.

Samuelson, P.A. and Nordhaus, W.D. (2001) *Economics* (17th edition). New York: McGraw-Hill.

Saraceno, E. (2004) Rural Development Policies and the Second Pillar of the Common Agricultural Policy, paper at the 87th EAAE-seminar "Assessing rural development policies of the CAP", 21–23 April, Wien.

Scanlan, H. (2002) Retirement scheme needs fine tuning. *Irish Farmers' Journal* 54(5), 2nd February 2002

Schindegger, F. and Tatzberger, G. (2002) *Polyzentrismus. Ein europäisches Leitbild für die räumliche Entwicklung.* Wien: Österreichisches Institut für Raumplanung.

Schramek, J., Biehl, D., Buller, H. and Wilson, G. (eds) (1999) *Implementation and Effectiveness of Agri-Environmental Schemes established under Regulation 2078/92,* Final Report on FAIR1 Project CT95-274, Frankfurt.

Schurmann, C. and Talaat, A. (2000) *Towards a European Peripherality Index,* Final Report. Institut für Raumplanung, Dortmund. Report for DG Regio. http://europa.eu.int/comm/regional_policy/sources/docgener/studies/pdf/periph.pdf

Scott, M. (2004) Building institutional capacity in rural Northern Ireland: the role of partnership governance in the LEADER II programme. *Journal of Rural Studies* 20, 49–59.

Scottish Affairs Committee (1996) The Future for Scottish Agriculture, Volume 1, Report together with the proceedings of the Committee, HMSO: London.

Scottish Office (1997) *Towards a Rural Development Strategy.* Edinburgh: The Scottish Office.

Shucksmith, M. (2003) *Territorial Aspects of the Common Agricultural Policy,* paper to European Congress of Rural Sociology, Sligo, August 2003.

Shucksmith, M. *et al.* (2004) *The Territorial Impact of CAP and Rural Development Policy,* final report of ESPON Project 2.1.3. Arkleton Institute, University of Aberdeen. http://www.espon.lu.213 FR

Sociologia Ruralis (2000) volume theme: Rural Development in Europe: the EU LEADER Programme Reconsidered.

Sotte, F. (2003) *An evolutionary approach to rural development,* paper to Conference of the Slovenian Associations of Agricultural Economists, Ljubljana.

Spathis, P. and Kaldis, P. (2003) Exploring factors conducive to maintaining Greek Farmers in LFAs, Paper presented at the Conference 'Less Favoured Areas and Development Strategies'. University of the Aegean, 20–23 November, Lesvos (in Greek)

SPSS (2000) Advanced Statistical Analysis using SPSS. Chicago IL: SPSS Inc.

Stapleton, L., Lehane, M. and Toner, P. (2000) *Ireland's Environment – a Millennium Report.* Wexford: Environmental Protection Agency

Statens Landbruksforvaltning (2003a) *Evaluering av tidligpensjonsordningen (Evaluation of the early retirement pension scheme).* Oslo: Norwegian Agricultural Authority (in Norwegian).

Statens Landbruksforvaltning (2003b) *Velferdsordninger – endringer som følge av jordbruksoppgjøret. Rundskriv nr. 063/2003.* Oslo: Norwegian Agricultural Authority (in Norwegian). http://www.slf.dep.no/index.asp?strUrl=1000784i&topExpand=1000021&subExpand=1000025&sub3Expand=&sub4Expand=

Statistisches Bundesamt (http://www.bml.de/landwirtschaft/ab-1999/material/tab008.htm)

Stubberud, K.V. and Samseth, K. (2000) Slekter kommer, slekter går – om generasjonskifte i landbruket (The times are changing – on the transfer of farms to the younger generation). NILF-rapport 2000:4. Oslo: Norwegian Agricultural Economics Research Institute. (in Norwegian).

Swinbank, A. and Tranter, R. (2004) *A Bond Scheme for Common Agricultural Policy Reform.* Wallingford: CABI Publishing.

Swinnen, J. (2003) *The EU Budget, Enlargement and Reform of the CAP and the Structural Funds,* paper to LUPG conference on Future Policies for Rural Europe 2006 and Beyond, Brussels, March 2003. See www.lupg.org.uk.

Tamme, O. and Dax, T. (2004) *Compensatory Allowances For Less-Favoured Areas, Synthesis Report.* Wien: Bundesanstalt fuer Bergbauernfragen.

Tamme, O., Bacher, L., Dax, T., Hovorka, G., Krammer, J. and Wirth, M. (2002) *Der Neue Berghöfekataster, Ein betriebsindividuelles Erschwernisfeststellungs-system in Österreich*, Facts & Features no. 23, Bundesanstalt fuer Bergbauernfragen, Wien.

Tanic, S. (ed.) (2001) *The Impact of Structural Adjustment Programmes on Family Farms in Central and Eastern Europe*, proceedings of FAO Expert Consultation, 20–23 January 2000, Budapest. http://www.fao.org/regional/SEUR/ExCon_Proceedin.pdf

Teagasc (2003) *REPS in a Changing Environment*, Proceedings of National REPS Conference. Dublin: Teagasc.

Terluin, I. (2001) Rural Regions in the EU – Exploring Differences in Economic Development, Groningen. Netherlands Geographical Studies no. 289.

Terluin, I.J. (2003) Differences in economic development in rural regions of advanced countries: an overview and critical analysis of theories. *Journal of Rural Studies* 19, 327–344

Terluin, I. and Venema, G. (2003) *Towards Regional Differentiation of Rural Development Policy in the EU*. The Hague: Agricultural Economics Research Institute, (LEI).

Terluin, I.J., Godeschalk, F.E., Von Meyer, H. and Strijker, D. (1995) Agricultural Income in Less Favoured Areas of the EC: A Regional Approach. *Journal of Rural Studies* 11(2), 217–228.

Tracy, M. (1997) *Government and Agriculture in Western Europe 1880–1988* (2nd ed.). Hemel Hempstead: Harvester Wheatsheaf.

Troeger-Weiss, G. (1998) *Regionalmanagement. Ein neues Instrument der Landes- und Regionalplanung.* Nr. 2 in Schriften zur Raumordnung und Landesplanung. Selbstverlag Fachgebiet Raumordnung und Landesplanung der Universitaet Augsburg, Augsburg.

University of Aberdeen Department of Agriculture and Forestry, and Macaulay Land Use Research Institute (2001) *Agriculture's Contribution to Scottish Society, Economy and Environment*, A Literature Review for the Scottish Executive Rural Affairs Department and CRU.

University of Bonn, Institute for Agricultural Policy 2004. CAPRI-DynaSpat. http://www.agp.uni-bonn.de/agpo/rsrch/dynaspat/dynaspat_e.htm

Van der Ploeg, J.D. (2003) Rural development and the mobilisation of local actors, paper at the 2nd Rural Development Conference in Salzbourg, 12–14 November.

Van Depoele, L. (2003) From sectorial to territorial based policies: the case of LEADER, *The Future of Rural Policy*, conference proceedings, OECD Headquarters, Paris.

van Tongeren, F. (2004) *Brief Review of the CAPRI Modelling System.* Prepared for the final project meeting of the CAP-STRAT project (QLTR-2000-00394, Louvain-la Neuve, 20 February 2004. LEI. http://www.agp.uni-bonn.de/agpo/rsrch/dynaspat/capri_review.pdf.

Ventura, F., Milone, P., Sabelli, D. and Autiello, L. (2001) Agri-tourism in Umbria: new on-farm activities and services, working paper for the research programme "The Socio-economic Impact of Rural Development policies: Realities & potentials" (FAIR CT 98-4288), University of Perugia.

Vindel, B. (2004) Implementation of EU rural development regulation in France: insights from the mid-term evaluation, paper at the 87th EAAE-seminar "Assessing rural development policies of the CAP", 21–23 April, Wien.

Voegel, F.T.R. (1993) Gute Argumente: Oekologische Landwirtschaft.Beck, Muenchen: Beck

Wallace, M. and Moss, J.E. (2002) Farmer Decision-Making with Conflicting Goals: A Recursive Strategic Programming Analysis. *Journal of Agricultural Economics* 53(1), 82–100.

Walsh, J. (1997) *Innovation in LEADER 11*, monograph Department of Geography, NUI Maynooth.

Walsh, J. (1999a) A strategy for sustainable local development. *Local Enterprise on the North Atlantic Margin*, (eds R. Byron and J. Hutson), 119–140. Aldershot: Ashgate.

Walsh, J. (1999b) The Ballyhoura 'Country' Local Development Model, *Best Practices in Local Development*, Paris. OECD.

Walsh, J. and Meldon, J. (2004) 2004 Urban Rural Relations in Ireland – case study of Ireland, prepared for ESPON Project 112 on Urban Rural Relations in Europe

Walsh. J.A., Brendan Kearney Associates and Fitzpatrick Associates (2000) *Evaluation of LEADER II Programme in Ireland* Report to Department of Agriculture, Food and Rural Development, Dublin, 130 pp.

Ward, N. (1993) The agricultural treadmill and the rural environment in the post-productivist era, *Sociologia Ruralis* 33, 348–364.

Wilkinson, A. and Korakas, A. (2001) SAPARD (Special Accession Programme for Agriculture and Rural Development), proceedings of conference "The Enlargement Process and the Three Pre-Accession Instruments", DG Enlargement, 5 March, Brussels.

Williams, F. (ed.) (2004) *Design and use of rural development measures in Scotland*, special study for the Scottish Executive Environment and Rural Affairs Department.

Williamson, O.E. (1996) *The Mechanisms of Governance*. Oxford: Oxford University Press.

Winter, M. and Gaskell, P. with Gasson, R. and Short, C. (1998) *The Effects of the 1992 Reform of the Common Agricultural Policy on the Countryside of Great Britain* (three volumes). Cheltenham: Countryside and Community Press and Countryside Commission.

WTO (2004) *Framework for Establishing Modalities in Agriculture*. Doha Development Agenda: Doha Work Programme, Annex A. Geneva: WTO.

WWF (2004) Implementing Rural Development Plans in Central and Eastern European Accession Countries, proceedings of a meeting Nov. 2003, Wien.

Zonneveld, W. (2000) Discoursive Aspects of Strategic Planning: a Deconstruction of the "Balanced Competitiveness" Concept in European Spatial Planning. *The Revival of Strategic Spatial Planning*. (eds W. Salet and A. Faludi). Amsterdam: Koninklijke Nederlandse Akademie van Wetenschappen.

Index